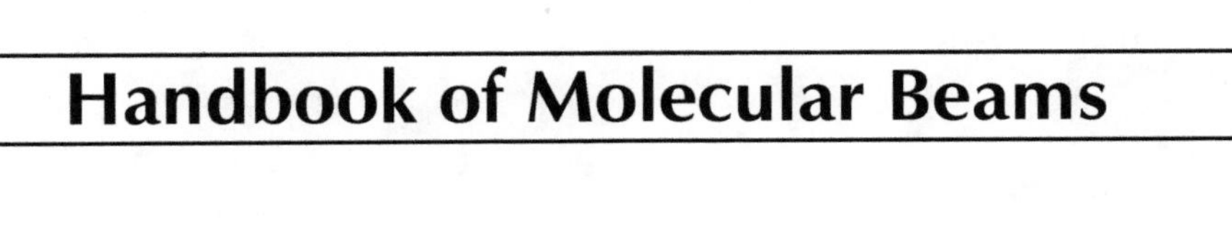

Handbook of Molecular Beams

Handbook of Molecular Beams

Editor: Josh Owen

www.murphy-moorepublishing.com

MURPHY & MOORE

Cataloging-in-publication Data

Handbook of molecular beams / Josh Owen.
p. cm.
Includes bibliographical references and index.
ISBN 978-1-63987-730-0
1. Molecular beams. 2. Molecular dynamics. I. Owen, Josh.
QC173.4.M65 H36 2023
539--dc23

Murphy & Moore Publishing
1 Rockefeller Plaza,
New York City,
NY 10020, USA

ISBN 978-1-63987-730-0

Contents

Preface

Every book is a source of knowledge and this one is no exception. The idea that led to the conceptualization of this book was the fact that the world is advancing rapidly; which makes it crucial to document the progress in every field. I am aware that a lot of data is already available, yet, there is a lot more to learn. Hence, I accepted the responsibility of editing this book and contributing my knowledge to the community.

A molecular beam refers to any type of ray or stream of molecules moving in the same direction, typically in a vacuum inside an evacuated chamber. It is created when a gas at a higher pressure is allowed to expand through a small orifice into a low-pressure chamber. This leads to the formation of a beam of particles, moving with equal velocities and few collisions. The molecules of a molecular beam are manipulated by the use of magnetic and electrical fields. These beams can be used to create thin films in molecular beam epitaxy. They are also used to fabricate structures such as quantum dots, quantum wells and quantum wires. This book elucidates the concepts and innovative models around prospective developments related to molecular beams. From theories to research to practical applications, studies related to all contemporary topics of relevance to these beams have been included herein. Those in search of information to further their knowledge will be greatly assisted by it.

While editing this book, I had multiple visions for it. Then I finally narrowed down to make every chapter a sole standing text explaining a particular topic, so that they can be used independently. However, the umbrella subject sinews them into a common theme. This makes the book a unique platform of knowledge.

I would like to give the major credit of this book to the experts from every corner of the world, who took the time to share their expertise with us. Also, I owe the completion of this book to the never-ending support of my family, who supported me throughout the project.

Editor

1

STIRAP: A Historical Perspective and some News

Klaas Bergmann

A very brief outline of what STIRAP is and does is followed by the presentation of the sequence of experiments, which started some 50 years ago, the visions developed and experimental efforts undertaken, that finally led to the development of STIRAP.

1 What Is STIRAP?

Stimulated Raman Adiabatic Passage (STIRAP, [1]) is a process which allows efficient and selective population transfer between discrete states of a quantum system, in its simplest form shown in Fig. 1. Level 1 is initially populated. The goal is to transfer all of that population to level 3. In most cases of interest, a direct one-photon dipole coupling between levels 1 and 3 is not possible. Therefore, one needs to invoke an intermediate level 2, often in a different electronic state. The characteristic and initially surprising feature of STIRAP is that the quantum system needs to be exposed first to the S-laser, which couples initially unpopulated levels. When the intensity of the S-laser is reduced, the intensity of the P-laser, which provides the coupling to the populated level, rises. If the switching-off of the S-laser and the switching-on of the P-laser is properly coordinated and the so-called adiabatic condition is fulfilled [2] nearly 100% of the initial population in level 1 will reach the target level 3 without ever establishing significant population in level 2. The underlying physics is interference of transition amplitudes, which—in the adiabatic limit—prevents population in level 2. Therefore, loss of population during the transfer process through spontaneous emission does not occur or is much reduced.

K. Bergmann (✉)
Fachbereich Physik der Technischen Universität Kaiserslautern, Erwin Schrödinger Str. 49, 67663 Kaiserslautern, Germany
e-mail: bergmann@rhrk.uni-kl.de

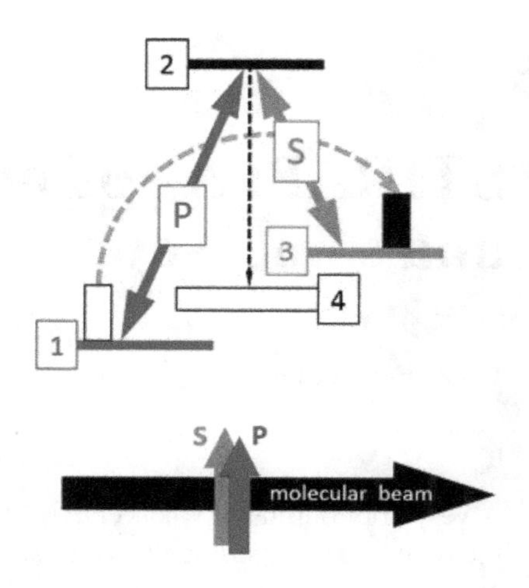

Fig. 1 The upper part shows a three-level system with the S- and P-lasers, coupling the levels 2–3 and 1–2, respectively. Level 4 stands for all levels that can also be reached by radiation from level 2. The lower part shows the geometric arrangement (S-before-P) for population transfer within particles of a molecular beam

The S-before-P sequence, called "counter-intuitive pulse sequence" in the early days, can be implemented either by suitably delayed laser pulses when applied to molecules in a gas cell (Fig. 2), or by spatially shifting the parallel axes of continuous lasers when population transfer within particles of a molecular beam is to be realized, as shown in the lower part of Fig. 1. It is the directional flow, which guarantees that a given molecule experiences a time variation of the coupling between levels as shown in the upper part of Fig. 2.

Another important feature, namely the robustness of STIRAP, made the scheme popular in many laboratories for applications in a wide and diverse range of quantum systems (see Sect. 6). Robustness means that a small variation of the S- or P-laser intensities or their time-delay does not reduce the transfer efficiency.

The original publication, reporting the main features of STIRAP and its theoretical foundation [1], was followed over the years by a number of review articles, e.g. [3–6]. The wide range of applications is documented in [7].

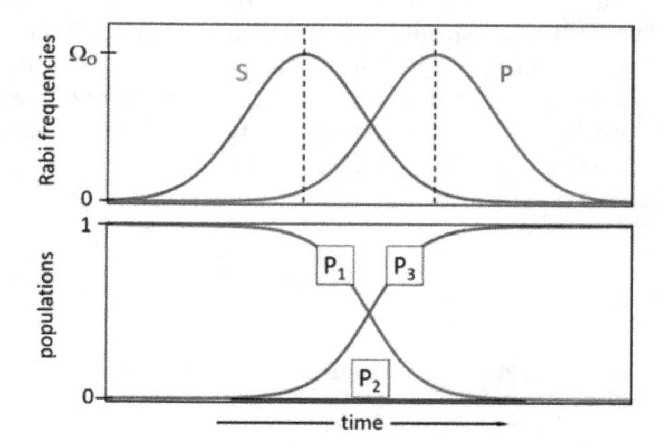

Fig. 2 The upper part shows the STIRAP-sequence of laser interactions with the quantum system (S-before-P). The variation of the Rabi frequencies, which determine the coupling strength between levels, is shown. The lower part shows the corresponding flow of population P_X from level 1 to level 3 (see Fig. 1). In the adiabatic limit no population is deposited in level 2

This article does not offer a detailed discussion of the physics of STIRAP. It describes, in the format of a memoir, the background, the vision, the various steps and the systematic plan followed, which finally led to realizing how a complete and robust population transfer between quantum states can be achieved. This work concludes with the presentation of a short list of topics or problems which benefited from the application of STIRAP. Although STIRAP has also been applied to many types of quantum systems, including a polyatomic molecule, the specific discussions and comments that follow relate mainly to diatomic molecules.

2 Background and Motivation

The deep roots of STIRAP reach back in time more than 50 years. The topic of my diploma thesis (submitted in early 1968) was the dynamics of photodissociation of some polyatomic molecules, using a classical pulsed high pressure discharge source [8]. After completion of that work my response to the question whether I wanted to continue this kind of experiments was a determined "no". I stated the reason: such work would very soon be done with lasers. In 1968 lasers were known for only 8 years.

Lasers did indeed play a central role in my PhD thesis, completed in early 1972. That work led to one of the very first applications of lasers to collision dynamics. The topic of the thesis emerged from spectroscopic work in sodium beams done by W. Demtröder while visiting R. N. Zare in Boulder [9]. In my work home-built Argon-ion lasers were used to excite a single rovibronic level (v´, j´) in the B-state of sodium molecules in a cell with rare gases added. (Here and below I use the traditional convention from spectroscopy: a single prime marks a level in an electronically excited state, while a double prime refers to a level in the electronic ground state.) Atom-molecule collisions induced transfer of population to neighboring rotational levels. That transfer was monitored by observing collision-induced spectral satellite lines. The pressure dependence of the intensity of those lines allowed the determination of rate constants. Of particular interest was the difference between rotational energy transfer to levels $(v', j' + \Delta j)$ and $(v', j' - \Delta j)$. The first paper on this topic [10] appeared in print only a few months after J. Steinfeld had published similar studies for I_2, also involving a laser [11]. Because the transferred energy was small compared to the mean kinetic energy, the observed difference of the rate constants for excitation and deexcitation processes with the same $|\Delta j|$ (called propensity) was unexpected. It was later explained through a detailed analysis of the wave functions involved [12].

While doing Ph.D. work, I learned about the then very popular molecular-beam technique through close contact with students working in a neighboring laboratory. The offer to continue academic work at the University of Kaiserslautern, founded in 1970, triggered the plan to combine molecular beams and lasers in future research. In early 1973, while carefully studying a paper by R. Drullinger and R. N. Zare on optical pumping of molecules [13], in particular their discussion of excitation and relaxation

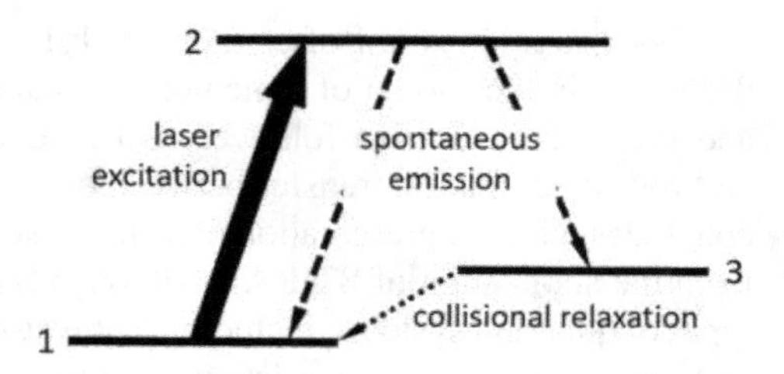

Fig. 3 The level scheme relevant for optical pumping of molecules in a gas cell showing laser-excitation, spontaneous emission and relaxation via collisions

pathways (see Fig. 3), I realized that the relaxation path after laser excitation and spontaneous emission back to the initially pumped level would be missing in the collision-free environment of a molecular beam. Thus the entire population of a specific thermally populated rotational level j''_{pump} could be removed. Controlled by Franck-Condon factors and optical selection rules, only a very small fraction of the laser-excited molecules would return to levels near j''_{pump} by spontaneous emission. This consideration led to the crossed beams arrangement as shown in Fig. 4. Particles scattered under the angle ϑ into the level j''_{probe} were probed by laser-induced fluorescence (see Fig. 5) while the pump laser would periodically switch off the population in level j''_{pump}. Most of the experiments involved levels in $v'' = 0$.

With the pump laser turned off, all thermally populated levels may contribute to the scattering into the probed level. With the pump laser turned on, the contribution from the pumped level would be missing. The difference of the scattering signal with pump laser off and on isolates the scattering rate from the level j''_{pump} into the level $j''_{probe} = j''_{pump} + \Delta j$ ($\Delta j > 0$ and $\Delta j < 0$ possible) under the scattering angle ϑ which is determined by the position of the narrow entry slit of a rotatable detector.

The molecular-beam laboratory for doing such experiments in Kaiserslautern (built after my post-doctoral work in Berkeley with C. B. Moore on laser-induced

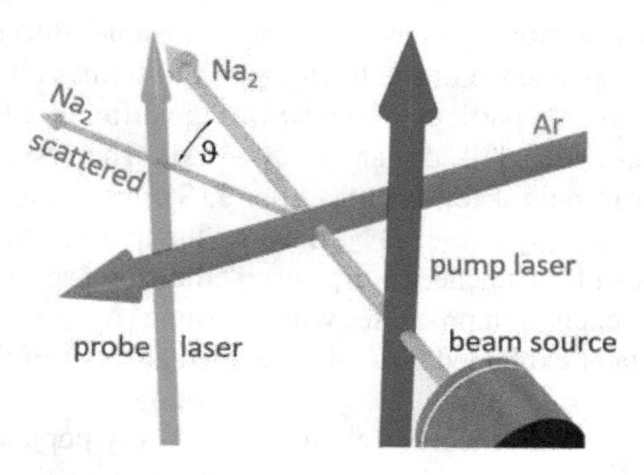

Fig. 4 The crossed beams arrangement for the study of state-to-state angle-resolved inelastic scattering, with a Na_2-beam, an Ar-beam, the pump-laser and a laser beam for monitoring the flux of particles scattered under the angle ϑ into the quantum state j''_{probe}. The device for collecting the fluorescence induced by the probe laser is not shown

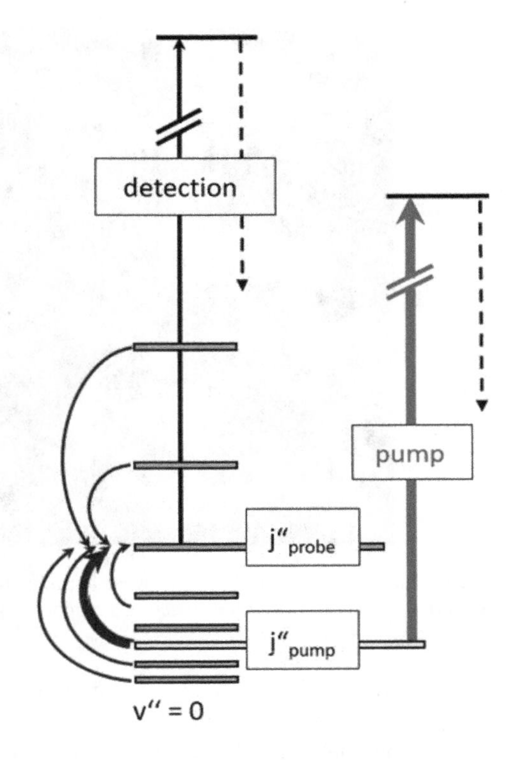

Fig. 5 The level scheme relevant for the scattering experiment as shown in Fig. 4. For further discussion, see the text

chemistry [14, 15]) had several innovative features. The entire apparatus was designed from the very beginning with the central role of lasers in mind. By design, some components, which were traditionally considered absolutely necessary, were not even included. In particular hot-wire detectors for detecting alkali atoms or molecules were replaced by lasers. Equally relevant: the mechanical flexibility of the detector, required for measuring angular distributions, was provided through the use of single-mode optical fibers in combination with a new design for efficient collection of laser-induced fluorescence [16]. A prerequisite of this work was a careful state-resolved characterization of the molecular beam [17].

Figure 6 shows what is most likely the very first AMO research laboratory with an optical-fiber network implemented, only a few years after Corning had manufactured the first Germanium doped single mode fibers. The photo shown was taken in 1977. Several lasers were connected to a number of experimental stations in different rooms with single-mode optical fibers donated by the fiber-research laboratory of Schott/Mainz. It was the late colleague Walter Heinlein from the electric engineering department of my university who introduced me to the relevant researchers in that laboratory. None of the optical components needed for coupling laser radiation into and out of fibers were commercially available. The photo was first shown in public at the ICPEAC conference 1979 in Kyoto. This photo triggered much more interest than the content of the related scientific presentation, namely the first laser-based

Fig. 6 The photo from 1977 shows what is probably the first laboratory with an optical fiber network installed, for flexibly connecting a number of lasers with various experimental stations

angularly resolved state-to-state energy transfer cross section [18]. The laser-based approach to molecular-beam scattering proved very successful. It led to a series of experiments yielding fully resolved state-to-state differential energy-transfer cross sections, with "rotational rainbows" being a prominent feature, see e.g. [19, 20], including even m-selectivity [21].

3 The Vision and the Challenge

Motivated by the success of the work mentioned in Sect. 2, I was considering in the late 1970s to use in my future work laser-based molecular state selection for laboratory studies of collision processes with relevance to atmospheric chemistry. It was known that chemical reactions and photodissociation processes in the higher atmosphere may lead to highly vibrationally excited molecules. However, little was known about how such vibrational excitation would change reaction-rates. The problem, though, was that the state-selection by optical pumping as described above works only for thermally populated levels. In a molecular-beam environment the levels $v'' \gg 1$ of interest are not populated. In order to use an approach similar to the one of Sect. 2 and shown in Fig. 4, one needs to efficiently and selectively populate a single rotational level in a highly vibrationally excited level. "Efficiency" was needed to realize a sufficiently high flux of excited molecules for scattering studies. "Selectivity" was crucial because reaction rates may sensitively depend on the vibrational level.

Further considerations, which are documented in my Habilitation Thesis submitted in late 1979 [22], led quickly to the conclusion that a direct one- or multi-photon excitation of a $v'' \gg 1$ level in the molecular electronic ground state would not be possible. The transition moments for high-overtone excitation are too small. Extremely strong laser pulses would be needed. At such high laser intensity many detrimental multi-photon excitation and ionization paths would make the approach inefficient. For homonuclear molecules the relevant transition moments are zero anyway.

Thus, it was straightforward to conclude that any efficient transfer scheme must invoke an auxiliary third level, most likely in an electronically excited state. On- or off-resonance Raman scattering, π-pulses or a sequential chirped adiabatic passage process were candidates. However, any process that drives the population through a level in an electronically excited state would suffer from unavoidable loss of population through spontaneous emission. Such processes would not only reduce the transfer efficiency to the target level; even worse, spontaneous emission would spoil selectivity because levels adjacent to the target level would also be populated. Off-resonance Raman scattering would reduce or avoid spontaneous emission, but the optical selection rules would prevent reaching levels $v'' \gg 1$ from thermally populated rotational levels in $v'' = 0$. Neither would the use of π-pulses allow reaching the goal, because the pulse area of the radiative interaction (roughly speaking: the product of the mean Rabi frequency $\Omega = \mu E/\hbar$ and the pulse duration, with μ being the transition dipole moment and E the electric field of the laser radiation) needs to be precisely controlled. Such transfer would not be robust. The transfer efficiency would depend very sensitively on small changes of relevant parameters. It would be different for different m-states within the rotational level. A sequence of two π-pulses would therefore allow the transfer of only a small fraction of the population of molecules in a given j" level, also because the condition for efficient transfer would be satisfied only for a tightly restricted number of trajectories of the molecule across the laser beam. A flux of molecules sufficiently high for scattering experiments would not be achievable.

Soon after reaching these conclusions the process of stimulated emission pumping (SEP) was proposed [23], which turned out to be very powerful for some collision dynamics experiments [24]. However, in most cases, the transfer efficiency did not exceed 10%. Thus, the problem was temporarily put aside with the hope that a new idea or a new inspiration would come along.

4 An Intermediate Step: The Molecular Beam Laser

The new inspiration came through discussions with B. Wellegehausen and after reading his papers on optically pumped lasers with alkali dimers in a heat pipe (length of the order of 10 cm) serving as gain medium [25]. In that work it was shown that high gain can be realized for transitions from levels in electronically excited states (populated by laser-pumping from thermally populated levels) to many high lying vibrational levels (see Fig. 7) in the electronic ground state. For many such transitions lasing was observed with a power of the pump laser as low as 1 mW. My quick back-of-the-envelope calculation showed that the gain in a molecular beam about 1 cm downstream from the nozzle would be even larger, despite the small extension of only a few mm. The reason is that the population distribution over low lying rovibronic levels in a supersonic beam is characterized by a temperature on the order of 10 K [17] rather than the ≈750 K in a heat pipe.

That conclusion quickly led to a preliminary design of a cavity around the vacuum chamber supporting the molecular beam. Figure 8 shows the second generation of such a cavity. An essential difference between a molecular-beam laser and a heat pipe relates to the relaxation process of population reaching the lower laser level (level 3 in Fig. 7). Such relaxation, removing continuously population from the lower laser level, is needed to allow continuous laser operation. In a heat pipe relaxation is dominated by collisions. In the molecular-beam, however, it is the directional flow that continuously transports new molecules in low-lying thermally populated levels (level 1 in Fig. 7) into the region of the laser cavity and at the same time removes molecules in level 3 from the cavity. These latter molecules do not experience further collisions. They remain in the highly vibrationally excited level and are available for collision experiments. Rotating the birefringent filter allows choosing which vibrational level is populated.

It was a very crucial moment when Uli Hefter and Pat Jones tried for the first time to get the molecular-beam laser going. It rarely happens that such experiments work at the first try. In this case it did happen on July 28, 1981. The cavity was aligned

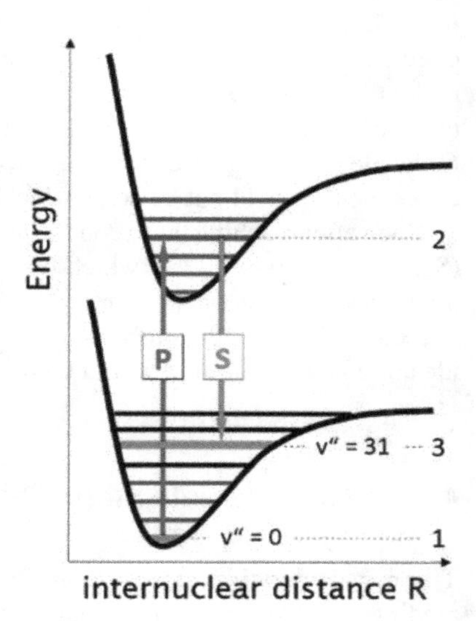

Fig. 7 The level scheme relevant for an optically pumped dimer laser, using molecules either in a heat-pipe or in a molecular beam

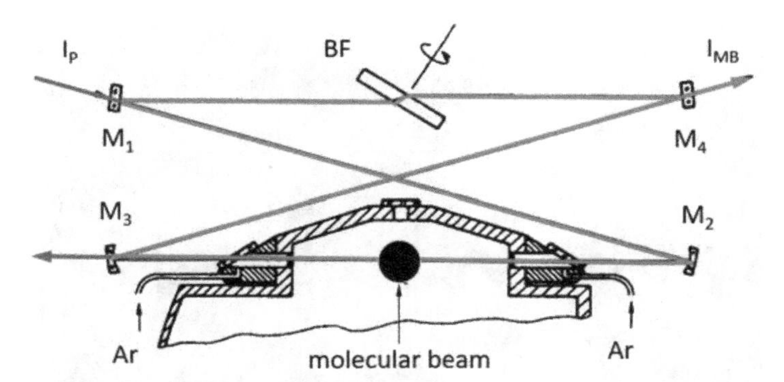

Fig. 8 The optical laser cavity built around the molecular beam in a vacuum chamber. I_P is the pump laser and I_{MB} the molecular-beam laser radiation. Only a small fraction of the pump laser is absorbed in the beam. Therefore, the pump laser exits again through the mirror M_3. BF is the birefringent (a tunable optical) filter that controls to which vibrational level v'' in the electronic ground state lasing is possible

and the molecular beam was operating. Upon turning on the pump laser lasing was immediately observed. The laser operated for about 15 min and then it went off. It took us 6 months (!!) to get it back into operation. This task was accomplished by Uli Gaubatz, who had joined the group shortly after the first operation of the laser was observed. Without these crucial 15 min, proving that the concept works, we would have probably given up after a few months and, most likely, the successful path towards realizing STIRAP would have been abandoned.

It turned out that one of the problems was the alkali deposit on the intra-cavity windows (see Figs. 8 and 9), separating the vacuum region from the outside. After several trial-and-error modifications, the problems were overcome. One of the measures was the heating of the windows to high temperatures what required the use of metal rings for vacuum-tight sealing. Furthermore, the installation of small pipes that directed a flow of Argon atoms away from the windows reduced the rate of deposit of alkali atoms and molecules on the windows. With these measures the molecular-beam laser could be routinely operated and in 1986 we set out to determine the properties of the transfer process, in particular its efficiency and selectivity [26].

The delay of a couple of years between the first demonstration of successful laser operation and the attempt to use such a laser for quantitative population transfer was in part due to the fact that no apparatus was available for such experiments. Students needed to first complete their ongoing experiments. The time was used to further explore the physics of the molecular-beam laser [27, 28]. Later we also built a molecular-beam laser with iodine molecules using a slit-nozzle expansion. For the latter system an optical pump-power threshold for starting the laser operation as low as 250 nW [29] was demonstrated.

Fig. 9 View through a window into the gain region of the molecular beam laser. The glowing red part is the molecular beam source with a cooler (darker) thermocouple attached. The thin yellow trace marks the pump laser beam which excites also background molecules. The thicker yellow region is seen because of radiation diffusion between particles along streamlines of the molecular beam. The gain region is the crossing between the two yellow traces

5 The Breakthrough

Figure 10 explains how the molecular-beam laser induced transfer efficiency was

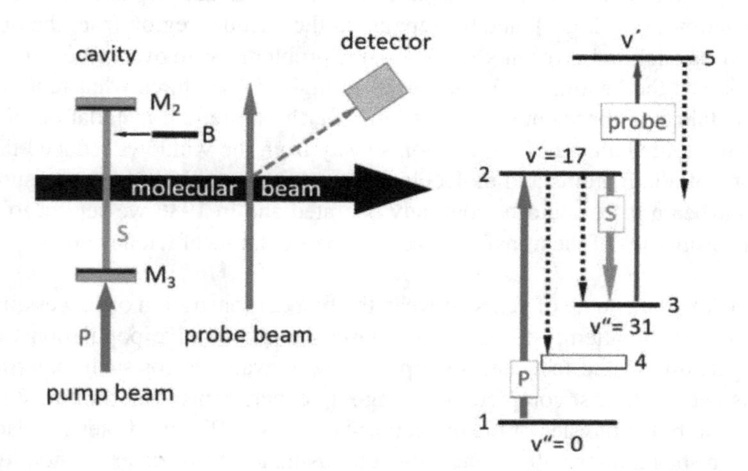

Fig. 10 Schematic of the set up (left part) for the calibration of the transfer efficiency induced by the molecular beam laser from level 1 ($v'' = 0$) to level 3 ($v'' = 31$). The element B is used to block the cavity. The related level scheme is shown in the right. Levels other than 1 and 3 that can also be reached from level 2 by spontaneous emission are summarized as level 4. The transfer into level 3 is probed by laser-induced fluorescence from level 5

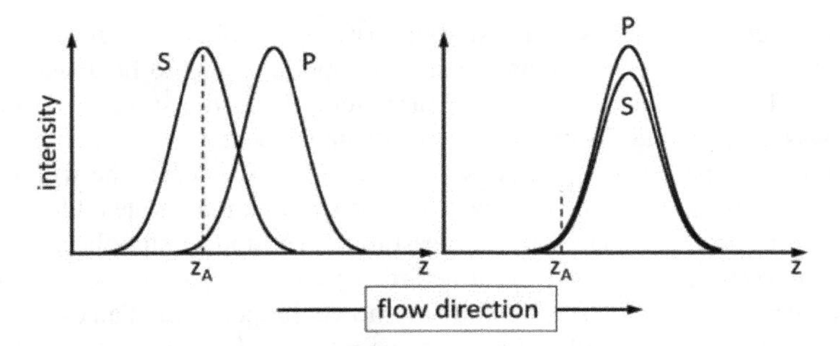

Fig. 11 The right part shows the P- and S- laser with coincident axes, as realized in the molecular-beam laser arrangement of Fig. 8. The left part shows the S-before-P arrangement of the lasers as also seen in the lower part of Fig. 1. The position z_A marks the location near which the pump laser would start pumping molecules out of level 1

determined. The pump laser excited molecules from the level $v'' = 0$ to $v' = 17$ in the B-state of Na_2, followed by spontaneous emission. The population in the target level of interest, e.g. $v'' = 31$, was probed by laser-induced fluorescence further downstream with the cavity blocked. Because the optical transition probabilities are known and after confirming that the population in level 1 was entirely depleted, the transfer efficiency of the population reaching level $v'' = 31$ by spontaneous emission could be determined. Unblocking the cavity allowed the molecular-beam laser to operate and the population in the target level increased. The increase of the population in relation to the known transfer efficiency by spontaneous emission yielded the beam-laser induced transfer efficiency. It was found [26] that the transfer efficiency was as large as 75% (larger than any other scheme would allow) but it was still far from the goal of $\approx$100%. The question thus was: what limits the transfer efficiency to about 75%?

The solution, which paved the final segment on the path to STIRAP, was surprisingly simple, as shown in Fig. 11. Results from earlier work on the consequences of optical pumping in two-step photoionization [30] led the way. The right part of Fig. 11 shows the profiles of the pump laser and the molecular-beam laser which appears after the pump laser is switched on. The axes of the two laser fields coincide. The molecules travel from left to right. As soon as the molecules reach the wings of the pump laser profile (P) they are efficiently pumped to the upper laser level. At the location z_A, however, the local molecular-beam laser intensity, which is supported by molecules that had already crossed the cavity, is still weak. Therefore, stimulated emission induced by the radiation field S that is supposed to populate level 3 (the target level) cannot yet compete with spontaneous emission. In fact, the transit time of the molecules across the cavity is about one order of magnitude longer than the radiative lifetime in level 2, the upper laser level. Therefore, only a fraction of the relevant molecules reaches the axis of the cavity where the beam laser is sufficiently strong to allow stimulated emission to compete successfully with spontaneous processes.

The final conclusion was again straightforward. Despite the significant effort invested to realize it, the molecular-beam laser approach had to be abandoned. In addition to the pump laser, an external laser for driving the stimulated emission process was needed, with its axis placed upstream of the axis of the pump laser, as shown in the left part of Fig. 11. As soon as the molecules enter the wings of the pump laser profile, they should be exposed to the maximum possible intensity of the S-Laser to optimize the chance for successful competition of stimulated emission with spontaneous processes. The informed reader realizes that the above argument, with optical pumping processes in mind, doesn't yet properly catch an essential part of STIRAP physics. It provided, however, the rationale for placing the axis of the S-beam upstream of the one for the P-beam, with a suitable overlap between the two.

It was very fortunate that Piotr Rudecki, a visiting scientist from Torun/Poland, arrived in the fall of 1987 a few days after I had come to the conclusion that a S-before-P arrangement was needed. The experiment which he wanted to join was not ready yet. However, he had some experience in modeling radiative processes. Therefore, I asked him to take our code for simulating the molecular-beam laser, which we had developed with the help of Wellegehausen, and modify it in accordance with the new geometry. The hope was to quantitatively understand the benefit of the S-before-P configuration from results of simulation studies. I certainly did expect a transfer efficiency of more than 75%.

About 10 days later, Rudecki presented his results: nearly 100% transfer. Expecting something near 90%, my reaction was: "hard to believe, please check for errors". A few days later, Piotr joined the group for the traditional after-lunch coffee-and-discussion meeting at a round table near the lab, presented the results of new calculations (see Fig. 12) and stated firmly "no errors—100% is correct". I

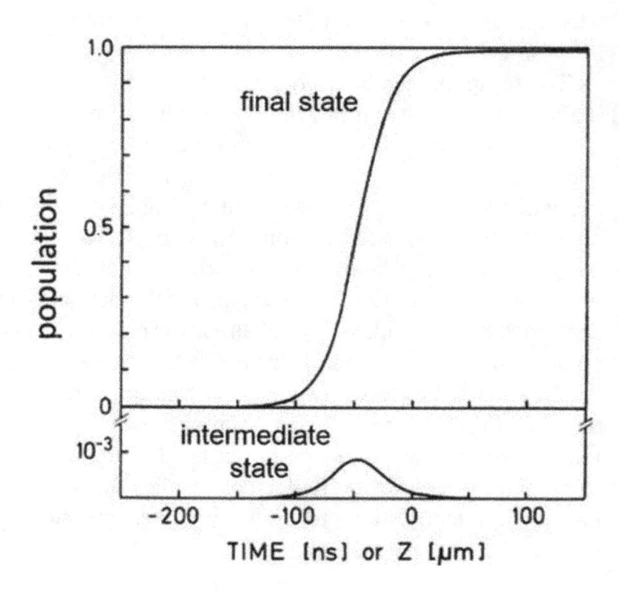

Fig. 12 The first numerical results for the transfer of the population of level 1 to level 3 (the final level) with a laser arrangement as shown in the lower part of Fig. 1. The lower part shows the transient population in the intermediate level 2. While, at late times, the final state population approaches unity, the maximal transient population of the intermediate level is 10^{-3}

clearly recall my prompt reaction: "congratulation—this result will be a big bang in the community and will determine what we do in the lab for the next 10 years".

A few months later, in early 1988, we managed to demonstrate that a very high population transfer can indeed be realized in the S-before-P configuration. In a first short publication [31], we showed results, but the theoretical basis had not yet been clearly sorted out. This gap was closed by the May 1, 1990 publication in J. Chem. Phys. [1], with Uli Gaubatz being the leading graduate student who had noted the close connection of what we did with the work by Claude Cohen-Tannoudji [32]. Prior to submitting the manuscript of paper [1], a detailed discussion of the adiabatic condition was published [2]. The paper [1] remains a standard reference for STIRAP and indeed, all the basic features that make STIRAP unique are experimentally documented and theoretically properly analyzed in that work. In that paper, also the acronym STIRAP was introduced. Regarding the latter, I had learned earlier that it is important to give a name to a new technique, method or process before others will do it. After the first rough draft of the paper was ready, I told my students that instead of joining the after-lunch meeting, I will spend one hour or two in my home-office and come back with a suggestion for an acronym. At home, I wrote down all physical processes or phenomena that had some connection to the transfer process, looked at the initial letter or letters and wrote them down in different orders. The criteria were: the acronym should have no more than two syllables and pronunciation should be easy. I returned to the lab with the suggestion STIRAP, which was accepted by all involved.

We defined the publication date or ref [1], May 1st, 1990 the "birthday" of STIRAP and celebrated its 25th anniversary in September 2015 with a well-attended and well-received international conference in Kaiserslautern [33].

6 Some STIRAP Highlights that Followed

The most recent compilation of some highlights regarding STIRAP applications can be found under [7]. Here a few topics are listed, with only one or two references given:

- preparation of ultracold molecules, see e.g. [34, 35]
- reduction of the upper limit of the electric dipole moment of the electron [36]
- controlling the phase of superposition states [37, 38]
- new tools for matter wave optics [39, 40]
- population transfer in superconducting circuits with relevance to quantum information [41, 42]
- single photon generation by sending atoms through an optical cavity [43]
- control of the pathway of light in optical fiber networks [44]
- population transfer in a solid-state environment [45, 46]
- controlled modification of the quantum state in strings of ions bound in a trap [47]

- use of the concept for acoustic waves with the potential to improve hearing aids [48]
- control of the flow of spin-waves in a network of suitable wave-guides [49]

The application of the STIRAP-approach to acoustic waves [48] is a particularly nice example for how far reaching the concept is. It also underlines that STIRAP is not a purely quantum mechanical process.

There are also a number of proposals with a detailed analysis of the feasibility of STIRAP applications, such as

- implementing quantum gates, e.g. [50–52]
- cooling particles in an atomic beam [53]
- excitation of molecular Rydberg states, by-passing predissociation levels [54]
- preparation of highly polarized molecular quantum states [55]
- storage of energy using nuclear isomers [56]
- spatial adiabatic passage (SAP, transfer of particles between traps) [57]
- digital pulse sequences, optimized via a learning algorithm, to speed up the process [58]

The example (SAP, in earlier years also called CTAP—coherent transport by adiabatic passage) would be a particularly intriguing demonstration of the STIRAP-concept: Consider three traps A, B and C in a linear arrangement and close proximity. Each trap is able to hold a single particle. Assume that one particle is initially in trap A while B and C are empty. When the coupling between the traps is properly varied as required for STIRAP (e.g. by lowering the barriers between them while keeping the quantum states in the traps in resonance) the particle is removed from A and appears in C without establishing a significant transient population in B.

7 Final Remarks

Following the original publication in 1990, the STIRAP concept has been systematically developed, both experimentally and theoretically, in Kaiserslautern (with too many publications to be all listed), also for applications beyond the canonical three-level system. That work benefitted greatly from the contributions of the visiting scientists Bruce W. Shore, Leonid P. Yatsenko, Razmik Unanyan, Matthew Fewell, and Nikolay V. Vitanov. Experimental progress was achieved through the dedicated work of many excellent students: Axel Kuhn, Stefan Schiemann, Jürgen Martin, Thomas Halfmann, Heiko Theuer, and Frank Vewinger to name at least some. The post-docs George Coulston, Horst-Günter Rubahn, and Stéphane Guérin also contributed significantly to the successful developments.

At the occasion of my first public presentation of the concept in the colloquium at JILA/Boulder on March 1st, 1990 (i.e. prior to the publication of [1]) I had the chance to discuss STIRAP with Peter Zoller. It was the follow-up theoretical and experimental work from the groups of Peter Zoller and Bill Phillips [59, 60], respectively, on matter-wave mirrors and beam splitters and the experimental work in the group of Steve Chu on atom interferometry [39], that made STIRAP quickly known in the AMO community. Nevertheless, it took more than 10 years after the original publication [1] before STIRAP was used in many laboratories and in different areas of research.

Several proposals did appear in the literature discussing the prospects for applying STIRAP to poly-atomic molecules. However, nearly all of them are based on model systems that did not adequately include relevant properties, such as the realistically modelled (detrimental) high level density. The consequences of the inclusion of these properties are carefully analyzed in an extensive simulation study [61] involving the HCN molecule. To the best of my knowledge, SO_2 is still the largest, or most complex, molecule to which STIRAP has been successfully applied [62] in an experiment.

As explained in Sect. 3, the STIRAP-concept was developed with reaction dynamics experiments involving vibrationally excited molecules in mind. One early experiment of that kind has been completed in Kaiserslautern ($Na_2(v'') + Cl \rightarrow NaCl + Na^*$ [63]). Using STIRAP was also essential in the recent observation of bimolecular reactions at ultracold temperatures [64]. However, the initial motivation had reactions of relevance to atmospheric chemistry in mind. Such an application still awaits its realization. Because of recent developments of coherent radiation sources for the region $\lambda < 200$ nm, this situation may change soon. The related requirements for the molecules H_2, N_2, O_2, and OH are discussed in the appendix of [5].

References

1. U. Gaubatz, P. Rudecki, S. Schiemann, K. Bergmann, Population transfer between molecular vibrational levels by stimulated Raman Scattering with partially overlapping laser: a new concept and experimental results. J. Chem. Phys. **92**, 5363–5376 (1990)
2. J.K. Kuklinski, U. Gaubatz, F.T. Hioe, K. Bergmann, Adiabatic population transfer in a three level system driven by delayed laser pulses. Phys. Rev. A **40**, 6741–6744 (1989)
3. K. Bergmann, H. Theuer, B.W. Shore, Coherent population transfer among quantum states of atoms and molecules. Rev. Mod. Phys. **70**, 1003–1026 (1998)
4. N.V. Vitanov, M. Fleischhauer, B.W. Shore, K. Bergmann, *Coherent Manipulation of Atoms and Molecules by Sequential Pulses*, in Advances of Atomic, Molecular, and Optical Physics, eds. by B. Bederson, H. Walther, vol 46 (Academic Press, 2001), pp. 55–190
5. K. Bergmann, N.V. Vitanov, B.W. Shore, Perspective—stimulated Raman adiabatic passage: the status after 25 years. J. Chem. Phys. **142**, 170901 (1–20) (2015)
6. N.V. Vitanov, A. Rangelov, B.W. Shore, K. Bergmann, Stimulated raman adiabatic passage in physics, chemistry and beyond. Rev. Mod. Phys. **89**, 015006 (1–66) (2017)
7. K. Bergmann (guest editor), Roadmap on STIRAP applications. J. Phys. B At. Mol. Opt. Phys. **52**, 202001 (1–55) (2019)
8. K. Bergmann, W. Demtröder, Mass-spectrometric investigation of the primary processes in the photo dissociation of 1,3-butadiene. J. Chem. Phys. **48**, 18–22 (1968)
9. W. Demtröder, M. McClintock, R.N. Zare, Spectroscopy of Na_2 using laser-induced fluorescence. J. Chem. Phys. **51**, 5495–5508 (1969)

10. K. Bergmann, W. Demtröder, Inelastic collision cross section for excited molecules: I. rotational energy transfer within the $B^1\Pi_u$-state of Na_2 induced by collision with He. Z. Phys. **243**, 1–13 (1971)
11. R.B. Kurzel, J.I. Steinfeld, Energy-transfer processes in monochromatically excited iodine molecules III. Quenching and multiquantum transfer from $v' = 43$. J. Chem. Phys. **53**, 3293–3303 (1970)
12. K. Bergmann, H. Klar, W. Schlecht, Asymmetries in collision induced rotational transitions. Chem. Phys. Lett. **12**, 522–525 (1972)
13. R.E. Drullinger, R.N. Zare, Optical pumping of molecules. J. Chem. Phys. **51**, 5523–5542 (1969)
14. K. Bergmann, C.B. Moore, Energy dependence and isotope effect for the total reaction rate of Cl + HI and Cl + HBr. J. Chem. Phys. **63**, 643–649 (1975)
15. K. Bergmann, S.R. Leone, C.B. Moore, Effect of reagent electronic excitation on the chemical reaction $Br(^2P_{1/2,3/2})$ + HI. J. Chem. Phys. **63**, 4161–4166 (1975)
16. K. Bergmann, R. Engelhardt, U. Hefter, J. Witt, A detector for state-resolved molecular beam experiments using optical fibers. J. Phys. E **12**, 507–514 (1979)
17. K. Bergmann, U. Hefter, P. Hering, Molecular-beam diagnostic with internal state selection: velocity distribution and dimer formation in a supersonic Na/Na_2 beam. Chem. Phys. **32**, 329–348 (1978)
18. K. Bergmann, R. Engelhardt, U. Hefter, P. Hering, J. Witt, State resolved differential cross sections for rotational transitions in Na_2 + Ne(He) collisions. Phys. Rev. Lett. **40**, 1446–1450 (1978)
19. U. Hefter, P.L. Jones, A. Mattheus, J. Witt, K. Bergmann, R. Schinke, Resolution of supernumerary rotational rainbows in Na_2-Ne scattering. Phys. Rev. Lett. **46**, 915–918 (1981)
20. G. Ziegler, M. Rädle, O. Pütz, K. Jung, H. Ehrhardt, K. Bergmann, Rotational rainbows in electron-molecule scattering. Phys. Rev. Lett. **58**, 2642–2645 (1987)
21. A. Mattheus, A. Fischer, G. Ziegler, E. Gottwald, K. Bergmann, Experimental proof of $|\Delta m| << j$ propensity rule in rotationally inelastic differential scattering. Phys. Rev. Lett. **56**, 712–715 (1986)
22. K. Bergmann, Molecular-beam experiments with internal state selection by laser optical pumping. Habilitation thesis (University of Kaiserslautern, 1979)
23. C. Kittrell, E. Abramson, J.L. Kinsey, S.A. McDonald, D.E. Reisner, R.W. Field, D.H. Katayama, Selective vibrational excitation by stimulated emission pumping. J. Chem. Phys. **75**, 2056–2059 (1981)
24. J.M. Price, J.A. Mack, C.A. Rogaski, A.M. Wodtke, Vibrational state-specific self-relaxation rate constant: measurements of highly vibrationally excited O2 (v = 19–28). Chem. Phys. **176**, 83–98 (1993)
25. B. Wellegehausen, Optically pumped cw Dimer laser. IEEE-QE **15**, 1108–1130 (1979)
26. M. Becker, U. Gaubatz, P.L. Jones, K. Bergmann, Efficient and selective population of high vibrational levels by near resonance Raman Scattering. J. Chem. Phys. **87**, 5064–5076 (1987)
27. P.L. Jones, U. Gaubatz, U. Hefter, B. Wellegehausen, K. Bergmann, Optically pumped sodium-dimer supersonic beam laser. Appl. Phys. Lett. **42**, 222–224 (1983)
28. P.L. Jones, U. Gaubatz, H. Bissantz, U. Hefter, I. Colomb de Daunant, K. Bergmann, Optically-pumped supersonic beam lasers: basic concept and results. J. Opt. Soc. Am. B **6**, 1386–1400 (1989)
29. I.C.M. Littler, S. Balle, K. Bergmann, Molecular beam Raman Laser with a 250 nW threshold pump power. Opt. Commun. **77**, 390–394 (1990)
30. K. Bergmann, E. Gottwald, Effect of optical pumping in two step photoionization of Na_2 in molecular beams. Chem. Phys. Lett. **78**, 515–519 (1981)
31. U. Gaubatz, P. Rudecki, M. Becker, S. Schiemann, M. Külz, K. Bergmann, Population switching between vibrational levels in molecular beams. Chem. Phys. Lett. **149**, 463–468 (1988)
32. C. Cohen-Tannoudji, S. Reynaud, Dressed-atom description of resonance fluorescence and absorption spectra of a multi-level atom in an intense laser beam. J. Phys. B **10**, 345–363 (1977)

33. K. Bergmann (conference chair), *Stimulated Raman Adiabatic Passage in Physics, Chemistry and Technology*. https://www.physik.uni-kl.de/bergmann/stirap-symposium-2015/
34. J.G. Danzl, M.J. Mark, E. Haller, M. Gustavsson, R. Hart, J. Aldegunde, J.M. Hutson, H.-C. Naegerl, An ultracold high-density sample of rovibronic ground-state molecules in an optical lattice. Nat. Phys. **6**, 265–270 (2010)
35. K.K. Ni, S. Ospelkaus, M.H.G. de Miranda, A. Pe'er, B. Neyenhuis, J.J. Zirbel, S. Kotochigova, P.S. Julienne, D.S. Jin, J. Ye, A high phase-space-density gas of polar molecules. Science **322**, 231–235 (2008)
36. V. Andreev, D.G. Ang, D. DeMille, J.M. Doyle, G. Gabrielse, J. Haefner, N.R. Hutzler, Z. Lasner, C. Meisenhelder, B.R. O'Leary, C.D. Panda, A.D. West, E.P. West, X. Wu, Improved limit on the electric dipole moment of the electron. Nature **562**, 355–364 (2018)
37. R.G. Unanyan, M. Fleischhauer, K. Bergmann, B.W. Shore, Robust creation and phase-sensitive probing of superposition states via stimulated Raman adiabatic passage (STIRAP) with degenerate dark states. Opt. Commun. **155**, 144–154 (1998)
38. F. Vewinger, B.W. Shore, K. Bergmann, *Superposition of Degenerated Atomic Quantum States: Preparation and Detection in Atomic Beams*, eds. by E. Arimondo, P.R. Berman, C.C. Lin. Advances in Atomic, Molecular and Optical Physics, vol. 58 (Academic Press, USA, 2010), pp. 113–172
39. M. Weitz, B. Young, S. Chu, Atomic interferometer based on adiabatic population transfer. Phys. Rev. Lett. **73**, 2563–2566 (1994)
40. H. Theuer, R.G. Unanyan, C. Habscheid, K. Klein, K. Bergmann, Novel laser-controlled variable matter wave beamsplitter. Opt. Expr. **4**, 77–83 (1999)
41. K.S. Kumar, A. Vepsäläinen, S. Danilin, G. S. Paraoanu, Stimulated Raman adiabatic passage in a three-level superconducting circuit. Nat. Commun. **7**, 10628 (1–6) (2016)
42. H.K. Zhu, C. Song, W.Y. Liu, G.M. Xue, F.F. Su, H. Deng, Ye Tian, D.N. Zhen, Siyuan Han, Y.P. Zhong, H. Wang, Yu-xi Liu, S.P. Zhao, *Coherent population transfer between uncoupled or weakly coupled states in ladder-type superconducting quitrits*. Nat. Commun. **7**, 11019 (1–6) (2016)
43. A.Kuhn, M. Hennrich, G. Rempe, Deterministic single-photon source for distributed quantum networking. Phys. Rev. Lett. **89**, 067901 (1–4) (2002)
44. S. Longhi, G. Della Valle, M. Ornigotti, P. Laporta, Coherent tunneling by adiabatic passage in an optical waveguide system. Phys. Rev. B. **76**, 201101 (1–4) 2007
45. J. Klein, F. Beil, T. Halfmann, Robust population transfer by stimulated Raman adiabatic passage in a Pr3 + :Y2SiO5 crystal. Phys. Rev. Lett. **99**, 113003 (1–4) (2007)
46. D.A. Golter, H. Wang, Optically driven rabi oscillations and adiabatic passage of single electron spins in diamond. Phys. Rev. Lett. **112**, (1–5) 116403 (2014)
47. J.L. Sørensen, D. Møller, T. Iversen, J.B. Thomsen, F. Jensen, P. Staanum, D. Voigt, M. Drewsen, Efficient coherent internal state transfer in trapped ions using stimulated Raman adiabatic passage. New J. Phys. **8**, 261 (1–10) (2006)
48. Y.X. Shen, Y.G. Peng, D.G. Zhao, X.C. Chen, J. Zhu, X.F. Zhu, One-way localized adiabatic passage in an acoustic system. Phys. Rev. Lett. **122,** 094501 (1–7) (2019)
49. P. Pirro, B. Hillebrands, The magnonic STIRAP process. J. Phys. B At. Mol. Opt. Phys. **52**, 202001 (22–23) (2019)
50. R.G. Unanyan, M. Fleischhauer, Geometric phase gate without dynamical phase. Phys. Rev. A **69**, 050302 (1–4) (2004)
51. D. Möller, J.L. Sörensen, J.B. Thomson, M. Drewsen, Efficient qubit detection using alkaline-earth-metal ions and a double stimulated Raman adiabatic passage. Phys. Rev. A **76**, 062321 (1–12) (2007)
52. B. Rousseaux, S. Guérin, N.V. Vitanov, Arbitrary qudit gates by adiabatic passage. Phys. Rev. A **87**, 032328 (1–4) (2013)
53. M.G. Raizen, D. Budker, S.M. Rochester, J. Narevicius, E. Narevicius, Magnetooptical cooling of atoms. Opt. Lett. **39**, 4502–4505 (2014)
54. T. J. Barnum, D. D. Grimes, R.W. Field, Populating Rydberg states of molecules by STIRAP. J. Phys. B At. Mol. Opt. Phys. **52**, 202001 (43–44) (2019)

55. S. Rochester, S. Pustelny, K. Szymanski, M. Raizen, M. Auzinsh, D. Budker, Efficient polarization of high-angular-momentum systems. Phys. Rev. A **94**, 043416 (1–12) (2016)
56. W.T. Liao, A. Pálffy, C. H. Keitel, A three-beam setup for coherently controlling nuclear state populations. Phys. Rev. C **87**, 054609 (1–12) (2013)
57. R. Menchon-Enrich, A. Benseny, V. Ahufinger, A.D. Greentree, T. Busch. J. Mompart, Spatial adiabatic passage: a review of recent progress. Rep. Prog. Phys. **79**, 074401 (1–31) (2016)
58. I. Poarelle, L. Moro, E. Prati, Digitally stimulated Raman passage by deep reinforcement learning. Phys. Lett. A **384**, 126266 (1–10) (2020)
59. P. Marte, P. Zoller, J.L. Hall, Coherent atomic mirrors and beam splitters by adiabatic passage in multilevel systems. Phys. Rev. A **44**, R4118–R4121 (1991)
60. L.S. Goldner, C. Gerz, R.J. Spreeuw, S.L. Rolston, C.I. Westbrook, W.D. Phillips, P. Marte, P. Zoller, Momentum transfer in laser-cooled cesium by adiabatic passage in a light field. Phys. Rev. Lett. **72**, 997–1000 (1994)
61. W. Jakubetz, Limitations of STIRAP-like population transfer in extended systems: The three-level system embedded in a web of background states. J. Chem. Phys. **137**, 224312 (1–16) (2012)
62. T. Halfmann, K. Bergmann, Coherent population transfer and dark resonances in SO_2. J. Chem. Phys. **104**, 7068–7072 (1996)
63. P. Dittmann, F.P. Pesl, J. Martin, G. Coulston, G.Z. He, K. Bergmann, The effect of vibrational excitation ($3 \leq v'' \leq 19$) on the chemiluminescent channel od the reaction Na2(v") + Cl $\rightarrow$ NaCl + Na*(3p). J. Chem. Phys. **97**, 9472–9475 (1992)
64. M.G. Hu, Y. Liu, D.D. Grimes, Y.-W. Lin, A.H. George, R. Vexlau, N. Bouloufa-Maafa, O. Dulieu, T. Rosenband, K.-K. Ni, Direct observation of bimolecular reactions of ultracold KRb molecules. Science **366**, 1111–1115 (2019)

2

Otto Stern and Wave-Particle Duality

J. Peter Toennies

Abstract The contributions of Otto Stern to the discovery of wave-particle duality of matter particles predicted by de Broglie are reviewed. After a short introduction to the early matter-vs-wave ideas about light, the events are highlighted which lead to de Broglie's idea that all particles, also massive particles, should exhibit wave behavior with a wavelength inversely proportional to their mass. The first confirming experimental evidence came for electrons from the diffraction experiments of Davisson and Germer and those of Thomson. The first demonstration for atoms, with three orders of magnitude smaller wave lengths, came from Otto Stern's laboratory shortly afterwards in 1929 in a remarkable *tour de force* experiment. After Stern's forced departure from Hamburg in 1933 it took more than 40 years to reach a similar level of experimental perfection as achieved then in Stern's laboratory. Today He atom diffraction is a powerful tool for studying the atomic and electronic structure and dynamics of surfaces. With the advent of nanotechnology nanoscopic transmission gratings have led to many new applications of matter waves in chemistry and physics, which are illustrated with a few examples and described in more detail in the following chapters.

1 Introduction

On September 8, 1926 Otto Stern submitted the first of a projected series of 30 articles all of which were to appear in the *Zeitschrift für Physik* under the subtitle "Untersuchungen zur Molekularstrahlmethode (UzM) aus dem Institut für physikalische Chemie der Hamburgischen Universität" [1]. In this initial article he outlined his plans for future physics experiments based on the method of molecular beams. It is remarkable that almost all of the projected experiments were successfully carried out in the ensuing 7 years until 1933 when Stern was forced to leave Germany. At the end of the list he had two special projects: "Der Einsteinsche Strahlungsdruck"

J. Peter Toennies (✉)
Max-Planck-Institut für Dynamik und Selbstorganisation, Am Fassberg 17, 37077 Göttingen, Germany
e-mail: jtoenni@gwdg.de

and "Die De Broglie-Wellen". Two years earlier, on November 1924, a little known young French physicist, Louis de Broglie, had defended his Sorbonne thesis [2]. In his thesis he published his famous formula $\lambda = h/mv$ [1] which introduced, for the first time, the concept of wave-particle duality of massive particles. Thus in 1926 Otto Stern was among the first who realized the tremendous importance of confirming de Broglie's revolutionary ideas by experiment.

After several marginally successful experiments, in May 1929 Otto Stern reported in a short note in the journal *Naturwissenschaften* well-resolved diffraction peaks in the reflection of He atoms from a cleaved NaCl crystal surfaces [3]. Thereby he provided the first direct evidence that the incident atoms of helium as massive particles had wave properties. Earlier in 1927, at about the same time as Stern and colleagues were still optimizing their experiment, Davisson and Germer had reported the first experimental evidence for the wave nature of electrons in *Nature* [4]. In the same year the wave nature of electrons was also independently confirmed in another *Nature* article by Thomson and Reid [5].

In the following, the history of the genesis of wave-particle duality will be reviewed. Then the diffraction experiments of Otto Stern will be described in more detail. Recent He atom diffraction experiments following on the footsteps of Stern and based partly on new developments are reviewed. These will serve as an introduction to the following articles describing the current state of experiments made possible by nanotechnologic advances which rely on the wave-particle duality of atoms and molecules.

2 History of Wave-Particle Duality and the de Broglie Relation

The early Greek philosophers Leucippus (fifth century BC) and his pupil Democritus (c. 460–c. 370 BC) were probably the first to introduce the idea that matter is composed of atoms (Greek for indivisible). Some of the early Greek philosophers also supposed that the *seen* object was emitting particles that bombarded their eyes. A modern theory that light was made of particles was first formulated much later by Isaac Newton (1643–1727). Somewhat earlier the idea that light was made of waves was postulated by Huygens (1629–1695). Soon after Thomas Young (1773–1829), August Jean Fresnel (1778–1827) and Joseph von Fraunhofer (1778–1826) carried out experiments which confirmed the wave nature of light. The wave versus particle dichotomy of light persisted up to the twentieth century. At the turn of the century, in 1900, Max Planck (1858–1947) introduced his famous radiation law and introduced the concept of a quantum of light which led the way to the development of Quantum Mechanics in 1926.

The modern development of the concept of matter waves can be traced back to 1905 when Einstein (1879–1955) used Planck's constant to explain the photoelectric

[1] λ is the wave length of a particle with mass m moving at velocity v. h is Planck's constant.

effect with the simple equation

$$KE_{el} = h\nu - \phi, \tag{1}$$

where KE_{el} is the kinetic energy of the ejected electron, ν is the frequency of the incident photons, and ϕ is the work function of the solid. Thereby it was established that photons act like particles with a fixed energy. Some years later, in 1923, Arthur Compton (1892–1962) reported that the X-ray photons that had been scattered from an electron in a solid, had a fixed momentum $P_{ph} = mc = \frac{h\nu}{c}$. Moreover, the momentum of the rebounding electrons could be explained by conservation of momentum and energy by assuming that the incident X-ray was a particle.

In 1919, these developments attracted the attention of Louis de Broglie (1892–1987) after he had been released from the army after World War I. Working in the laboratory of his physicist brother he became interested in the new concept of a "quanta of light". In 1922 he published two short articles on black-body radiation [6, 7]. Then, in 1923 he published three additional short two-page articles in *Comptes Rendus* [8] in which he developed his ideas on the wave nature of light. These he summarized in a half-page *Nature* article in 1923 [9] and in the Philosophical Magazine [10]. In the *Nature* article he concluded: "A radiation of frequency ν has to be considered as divided into atoms of light of very small internal mass $\left(< 10^{-50}\ \mathrm{gm}\right)$ which move with a velocity very nearly equal to c given by $\frac{m_0c^2}{\sqrt{1-\beta^2}} = h\nu$. The atom of light slides slowly upon the non-material wave the frequency of which is ν and velocity c/β, very little higher than c. The phase wave has a very great importance in determining the motion of any moving body, and I have been able to show that the stability conditions of the trajectories in Bohr's atom express that the wave is tuned with the length of the closed path." With this he anticipated his wave hypothesis in the case of the electron orbits in an atom by showing that the circumference of the orbits would be an integral multiple of the wavelength of the electron.

De Broglie's *These de doctoral* appeared in the following year 1924 and later was published in 1925 in *Annales de Physique* [11]. In only one place in the final short chapter in one of the last sections entitled *The New Conception of Gas Equilibrium* he writes his famous formula

$$\lambda = \frac{h}{mv}. \tag{2}$$

The fact that the formula appears only once suggests that at the time he was apparently more interested in discussing wave motion in general as applied to X-rays and electrons and did not fully realize then the far-reaching significance of the equation for which he is presently known.

According to the excellent reviews of de Broglie's discovery by Medicus [12] and MacKinnan [13] de Broglie's work did not become widely known, partly because *Comptes Rendus* was not very popular and partly because the reputation of the little-known young theoretician was controversial. De Broglie's thesis only became widely known after Paul Langevin, who was a member of his examination committee, had sent it to Einstein. Einstein immediately appreciated the far reaching consequences of de Broglie's ideas and wrote back that de Broglie had "lifted a corner of the great veil" and incorporated the new concept in his article in the *Proceedings of the Prussian Academy* which appeared in 1925 [14].

3 1925: Experimental Confirmation for Electrons

About the time of de Broglie's theories Clinton J. Davisson at the American Telephone and Telegraph (now AT&T) and the Western Electric Company in New York City was experimenting on the effect of electron bombardment on metal surfaces. This research was carried out in connection with understanding the physics of vacuum tubes which were a major product of the two companies. In 1921, Davisson and Kunsmann had reported their initial results on the measurements of the angular distributions of scattered electrons with incident energies up to 1000 eV in *Science* [15] and later in *Physical Review* [16]. At energies below 125 eV upon scattering from platinum and magnesium metal surfaces, they observed an unexpected small lobe in an otherwise Gauss-shaped distribution (Fig. 1). The lobe, they thought, could be related to the Bohr-model electron orbits in the metal.

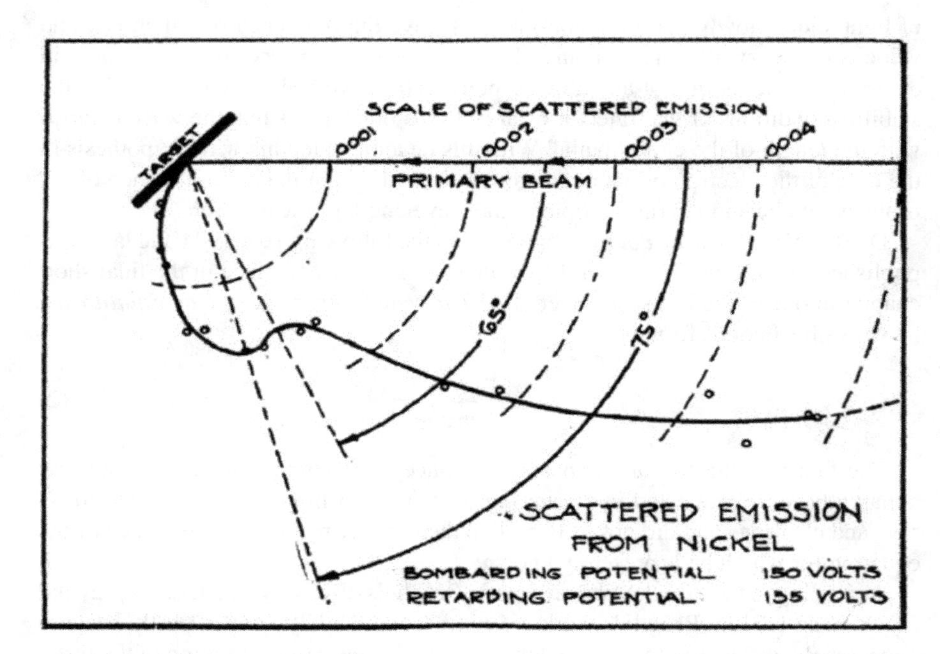

Fig. 1 The angular distribution of scattered electrons from nickel observed by Davisson and Kunsmann in 1921. The figure is taken from Ref. [16]

In the summer of 1925, in far away Göttingen, Friedrich Hund (1896–1997) gave a talk about the experiments of Davisson and Kunsmann in Prof. Max Born's seminar on *Die Struktur der Materie.* The Göttingen physicists were then also interested in electron scattering in connection with the 1924 Franck-Hertz experiment, which Franck and Hertz had carried out in Berlin before coming to Göttingen. In Göttingen it was one of the areas of research of the Born group. Walter Elsasser (1904–1942), a student attending the seminar, who had read about de Broglie's theory which had appeared in the same year, conjectured that the lobes reported by Davisson and Kunsman were in fact partly resolved diffraction peaks and could be the first experimental confirmation of de Broglie's theory. Since Elsasser was also attending the lecture course given by Prof. James Franck (1882–1964), he told Franck about his thoughts, whereupon Franck encouraged Elsasser to write a short article about his idea (Fig. 2). The half-page letter appeared in August 1925 in *Die Naturwissenschaften* [17].

The previous account is from the American National Academy of Sciences Biographical Memoir about Elsasser written by Rubin [18]. The important role of Elsasser is also supported by Hund [19] and also by Max Born in his article on the quantum mechanics of collisions [20]. Later, however, Max Born claimed that he had the idea and that he was the one to encourage Elsasser to write the note [21]. In the same article he does remark that he cannot fully remember the details. A somewhat

Fig. 2 The story behind the first realization of experimental evidence for wave-particle duality according to the official National Academy of Science biography of Walter Elsasser [18]

different story can be found in Gehrenberg [22, 23]. It is also interesting to note that Davisson and his assistant Germer were not at all convinced by Elsasser's idea. In their 1927 article in *Physical Review* [24] they wrote "We would like to agree with Elsasser in his interpretation of the small lobe reported by Davisson and Kunsman in 1921 and 1923, but are unable to do so". At the time they were convinced that the curves seen in their initial experiments were "unrelated to crystal structure".

At about the same time, in April 1925, a momentous accident occurred in the laboratory of Clinton Davisson. The glass vacuum tube containing the electron scattering apparatus exploded while the metal target was at high temperatures. In an attempt to save the highly oxidized target it was subsequently baked out over an extended period of time in an effort to reduce the oxide coating. When the experiments were repeated, surprisingly, much sharper lobes were observed, which were especially apparent when the crystal was rotated in the azimuthal direction (Fig. 3). In their *Nature* article in 1927 [4] Davisson and Germer attributed the six sharp azimuthal features to matter-wave diffraction in agreement with the theory of de Broglie. In this same article they do finally give credit to Elsasser. In the same year they published a complete analysis in Physical Review [24]. There they attributed the appearance of the newly found diffraction peaks to the increased order of the sample, which had been partly crystallized by the annealing during the long bake out [24]. With serendipitous good fortune, they had finally succeeded in providing convincing confirmation of the

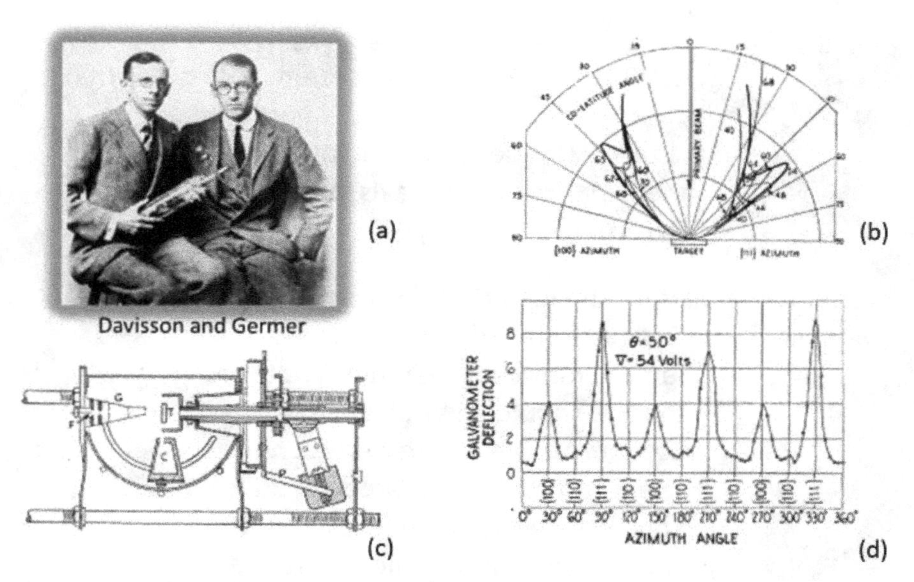

Fig. 3 The first unequivocal evidence for the de Broglie formula came from the 1927 electron diffraction experiments of Davisson and Germer (**a**). **b** The polar angle distribution shows diffraction peaks on both sides of the specularly reflected beam. **c** Schematic diagram of the electron scattering apparatus. **d** The sharp diffraction peaks observed on rotating the crystal. The latter 3 figures are from Ref. [24]

de Broglie relation. Of course this brief account does not in any way do justice to the many agonizing attempts that finally led to Davisson and Germer's accidental but successful experiment. The full story has been documented in detail by Gehrenberg [22].

This pioneering experiment was the forerunner of modern *Low Energy Electron Diffraction (LEED)* and *Electron Energy Loss Spectroscopy (EELS)* methods for surface analysis. Presently the former is widely used to determine the structure mostly of metal surfaces and the latter for measuring the surface phonons and the vibrations of clean and adsorbate-covered surfaces.

Then also in 1925, George P. Thomson (1892–1975), the son of the famous English physicist J. J. Thomson, reported on the diffraction of high energy electron beams (3,900–16,500 eV) upon passage through a 30 nm thick celluloid film. Their *Nature* article [8] appeared only two months after the *Nature* article of Davisson and Germer. With their simple but elegant experiment they were able to also verify the de Broglie relation (Fig. 4) [25]. In 1937, both Davisson and Thomson were awarded the Nobel Prize for *The Discovery of the Electron Waves.*

The significance of these developments has recently been highlighted in a thought provoking article by Steven Weinberg entitled *The Trouble with Quantum Mechanics* [27]. He begins his *critique* of quantum mechanics by noting "Then in the 1920s, according to theory of Louis de Broglie and Erwin Schrödinger, it appeared that

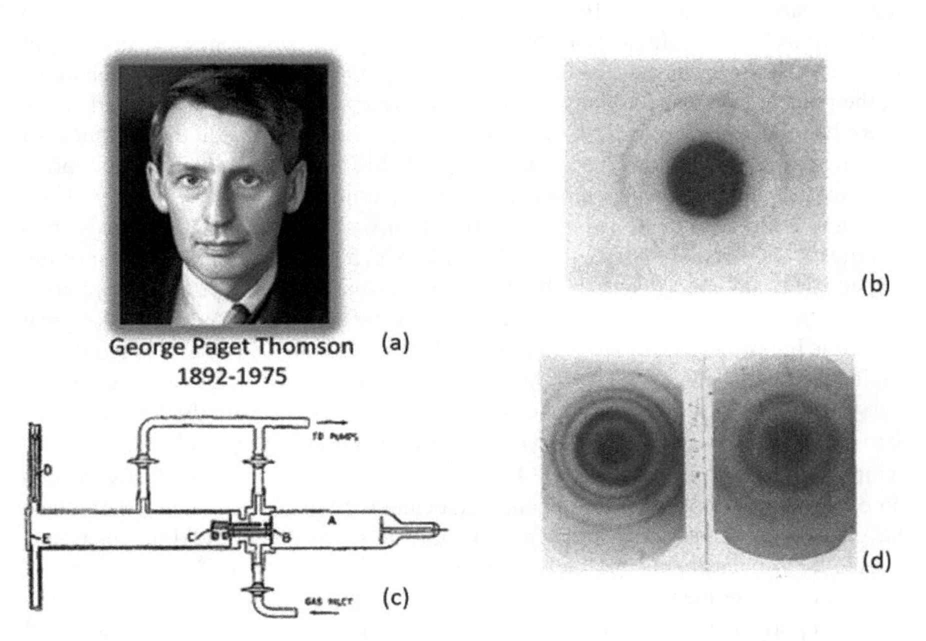

Fig. 4 The electron transmission experiment by G. P. Thomson which appeared a few months after the results of Davisson and Germer. **a** Photo of G. P. Thomson. **b** Diffraction rings on transmission through celluloid. **c** The electron transmission apparatus. **d** Diffraction rings seen on transmission through two different thin gold foils. The later 3 figures are from Ref. [26]

electrons which had always been recognized as particles, under some circumstances behaved as waves. In order to account for the energies of the stable states of atoms, physicists had to give up the notion that electrons in atoms are little Newtonian planets in orbit around the atomic nucleus. Electrons in atoms are better described as waves, fitting into an organ pipe. The world's categories had become all muddled."

4 Otto Stern's Experimental Confirmation for Atoms

Otto Stern's career as an experimentalist started in 1919 when he took up the position of assistant in Max Born's two-room theory group at the University of Frankfurt [28]. With a cleverly conceived apparatus he was able, for the first time, to measure the mean velocity in a molecular beam, which had been predicted by Clausius [29]. Then, in 1921 he embarked with Walter Gerlach on the famous Stern-Gerlach experiment [30], which led to the discovery of angular momenta and magnetic moments of atoms in magnetic fields. In the fall of the same year, Stern left Frankfurt to take up a new position as "Extraordinarius" (associate professor) at the University of Rostock. In the aftermath of World War I, the financial conditions in Rostock were such that he could not think of carrying out experimental research. Fortunately, Stern's Rostock period lasted only one year. In the following fall of 1922 he accepted an offer as an "Ordinarius" (full professor) of Physical Chemistry and Director of the Institute of Physical Chemistry at the University of Hamburg. On January 1, 1923, in the midst of the great inflation in Germany, he took up his research activities at Hamburg. Here Stern had the good fortune to be assigned four laboratory rooms in the basement of the Physics Institute. Now Stern was finally able to continue the experiments started in Frankfurt and to plan new molecular beam experiments.

Then, as already mentioned in the Introduction, on September 8, 1925 Otto Stern's manifesto appeared in *Zeitschrift für Physik* in which he outlined his plans for future molecular beam experiments including the confirmation of the de Broglie relation [19]. In this connection, he wrote (translated by the author) "...A question of great principle importance is the real existence of de Broglie waves, i.e. the question if with molecular beams, in analogy to light beams, diffraction and interference phenomena can be observed. Unfortunately, the wave lengths calculated with the formula of de Broglie ($\lambda = \frac{h}{m \cdot v}$), even under the most favorable conditions (small mass and low temperatures), are less than 1 Å (=10^{-1} nm). Nevertheless, the possibility in such an experiment to observe these phenomena cannot be excluded. Such experiments have so far not been successful." In a footnote he noted that at the time when they started their experiments in Hamburg he was not aware of the 1927 experiments of Davisson and Germer.

The report on the first experiments, which were judged publishable, was submitted on Christmas eve of 1928 as publication No. 11 in the series *Untersuchungen zur Molekularstrahlmethode* (UzM) with Friedrich Knauer [31]. In this initial experiment molecular beams of H_2 and He were scattered under grazing angles of only 10^{-3}

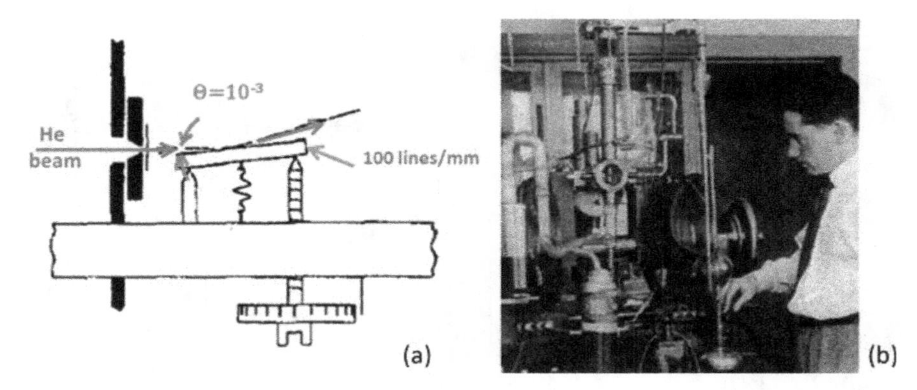

Fig. 5 The first experiment by Stern for a test of the de Broglie relation for massive particles. **a** A schematic diagram of the apparatus taken from the publication Ref. [31]. **b** A photo of a typical glass vacuum apparatus used by Stern and his colleagues

radians from a flat ruled optical grating made either of brass, glass or steel with up to 100 grooves per mm (Fig. 5). These experiments were only partially successful in the sense that only a sharp specular reflected beam but no diffracted beams were observed. The latter, it was concluded, were too close to the specular peak to be resolved. In the same article they describe a second apparatus which was designed for scattering at large angles. With this apparatus they scattered various atoms and molecules including H_2, He, Ne, Ar and CO_2 from a freshly cleaved and continually heated (100 C) NaCl crystal. At the time it was well established that the ionic alkali halide crystals were easily cleaved and that the resulting surfaces were relatively free of defects. Again, only for H_2 and He could relatively sharp specular scattering be observed. In the same UzM article they noted that parallel research was going on in the U.S. by Johnson [32] who had reported a specular peak with H atoms scattered from LiF and Ellet and Olsen [33] who scattered Cd and Hg atoms from NaCl, and that they had also been unsuccessful in observing diffraction.

Several months later on April 20, 1929, as mentioned in the Introduction, Otto Stern submitted a short note to the *Naturwissenschaften* reporting that with the second apparatus he had now found convincing evidence for the sought-after diffraction peaks with both He and H_2 in scattering from the surface of a single crystal of NaCl [6]. One year later, Immanuel Estermann and Otto Stern in UzM No. 15 reported that with an improved apparatus they were now able to observe well-resolved diffraction peaks both with H_2 and He from LiF, NaCl and KCl [34] (Fig. 6). The diffraction angles obeyed the de Broglie relation $\left(\lambda = \frac{h}{m \cdot v}\right)$ calculated from the lattice constants of the crystals and the masses of the scattering particles, both of which were well-known at the time.

Apparently, Otto Stern was still not completely satisfied judging by the fact that in the following year in 1930 he embarked on an ambitious project to rigorously and quantitatively check the de Broglie relation. For this it was necessary to use a beam with a well-defined velocity. In the following pioneering experiments two

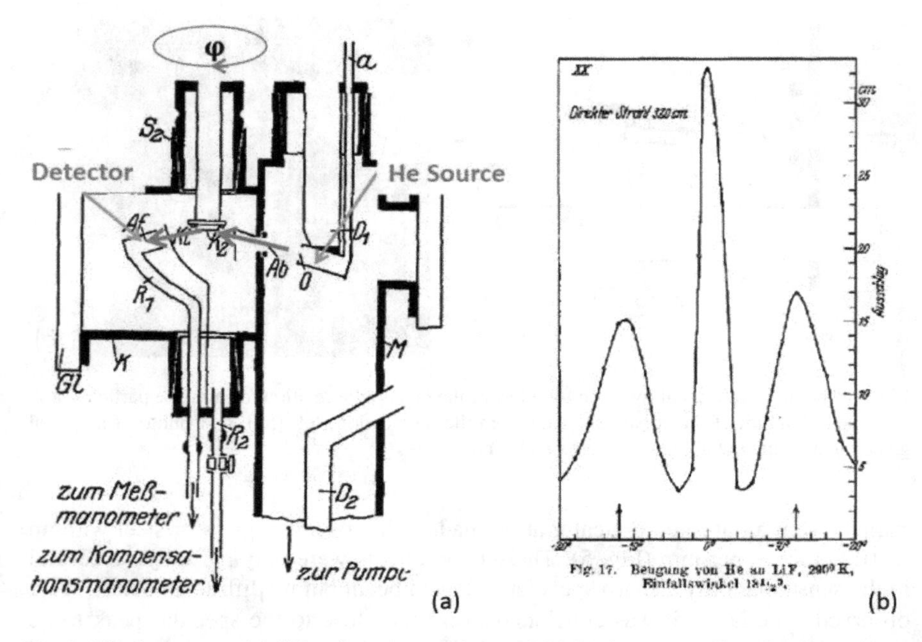

Fig. 6 The first successful diffraction of a massive particle, He from crystalline LiF observed by Stern in 1929 [3]. **a** The glass encased vacuum apparatus in which the crystal was rotated. **b** The diffraction pattern showing two first order diffracted peaks on both sides of the specular peak [34]

new methods were used to select velocities from the broad Maxwell-Boltzmann distribution [34]. One method exploited the dependence of the first order diffraction angle on the incident beam velocity. In this apparatus, the He atom beam was first diffracted from one crystal surface and only those atoms diffracted into a chosen solid angle, corresponding to the desired velocity, were transmitted. These atoms were then directed at the second crystal surface under investigation. Figure 7 shows a schematic of the method and a more detailed view of the actual apparatus.

The second method was based on two rotating slotted discs on a common axis with the second downstream disc rotationally displaced. The discs were displaced in such a way that only atoms transmitted by the slots of the first disc with the desired velocity could pass through the slots of the second disc. For the motor they modified a Gaede-Siegbahn molecular turbo pump which was already available at the time.[2] The 1 cm thick discs with the pump veins were replaced by two 12 cm dia 1 mm thick discs 3.1 cm apart each with 408 slits with a width of 0.4 mm.[3] (Fig. 8).

[2]The molecular turbo pump was invented by Gaede in 1910; improved by Holweck in 1923 and by Manne Siegbahn in 1926.

[3]In the operation of the two disc velocity selector Stern and colleagues appear to have neglected the velocity side bands. For the transmitted velocity they assumed that only those atoms that passed a slit in the first disc could pass the next slit in the second disc-on the same axis-but displaced rotationally by one period with respect to the first disc. Thus the velocity is given by $v_0 = \ell \cdot \nu \cdot N$, where ℓ is the distance between discs, which is divided by the time for the disc to move by one slit, $\tau = 1/\nu N$,

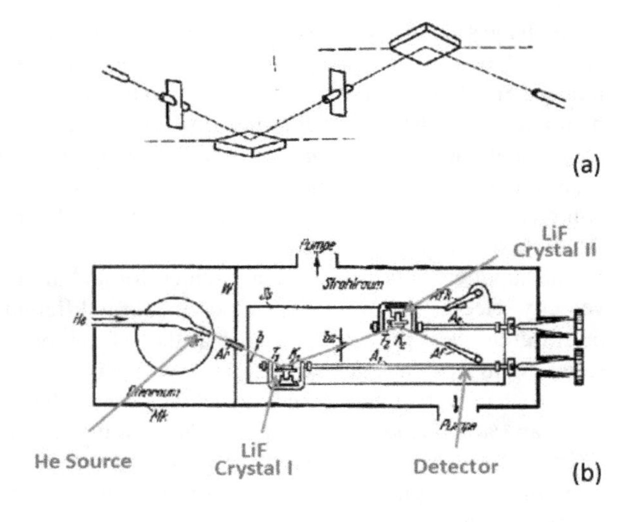

Fig. 7 The apparatus used by Estermann, Frisch and Stern to measure diffraction of a velocity selected He atom beam [34]. **a** Schematic of the apparatus. By choosing the diffraction angle only atoms with the desired velocity are transmitted to the second crystal for diffraction. **b** Side view of the apparatus

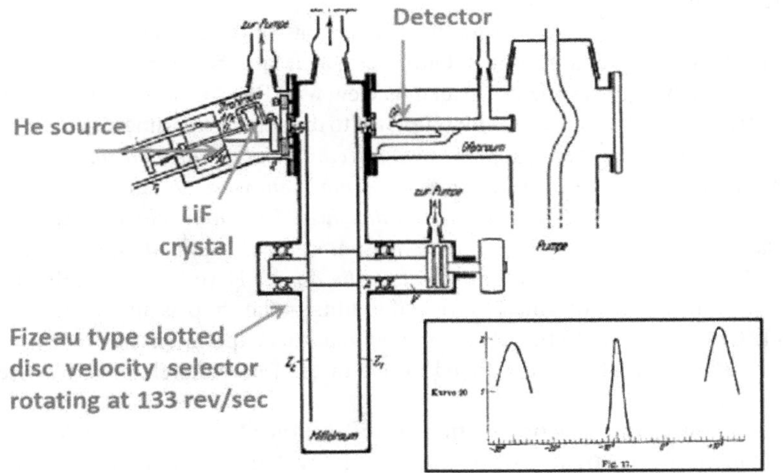

Fig. 8 The second apparatus used by the same authors as in Fig. 7 to measure the diffraction of a velocity selected He atom beam [34]. **a** The velocity is selected by two rotating identical slotted discs, rotational displaced. **b** The sharp diffraction peaks made it possible to confirm the de Broglie wavelength within less than 1%

---

here ν is the frequency of rotations and N is the number of slits. But atoms with slower velocities will arrive at a later time when the second disc has rotated to the position of the second slit with velocity $\upsilon_1 = \frac{1}{2}\upsilon_0$ and be transmitted. Also very fast atoms that pass the slit in the second disc in the same position as in the first disc will be transmitted. For this reason, modern velocity selectors have several additional discs at strategic distances along the axis to block out the unwanted velocity side bands. The contribution of these side bands in the case of the Stern experiments is estimated by the author to amount to about 25%.

With the second apparatus Immanuel Estermann, Robert Frisch, and Otto Stern in No. 18 of the UzM series reported highly resolved diffraction patterns of monochromatized He atoms from LiF [34] (Fig. 8). In a footnote of the same UzM article they mention that in the initial measurements the experimental wavelength was smaller by 3% than the predicted wavelength. This, they were convinced, was far too large a discrepancy to be possible. After a long search it was ultimately found that the commercial precision graduated disc used by the Hamburg machine shop to locate the slots to be milled in the discs had 408 positions instead of the 400 specified by the supplier! When this was accounted for and after a careful calibration of the velocity selector and extensive measurements at different rotational frequencies and on different days they established that the diffraction angle was 19.45 deg corresponding to a de Broglie wave length $\lambda = 0.600 \times 10^{-8}$ cm. The de Broglie wavelength calculated for the transmitted velocity was 0.604×10^{-8} cm with a deviation of less than 1% from the measured value. This was the first precision measurement of the de Broglie wavelength of a massive particle. Otto Stern's strong conviction that something must be wrong with the initial results illustrates once more his extraordinary acumen as an experimentalist.

After 4 years, Otto Stern had finally achieved one of the goals that he had laid out when he came to Hamburg. In fact, Otto Stern, as it later became clear, was still not satisfied with these experiments. In an interview with Res Jost in Zürich in 1961 [37] Otto Stern began his comments with reference to the above experiments: "*I especially like this experiment, it is not properly recognized. It is about the determination of the de Broglie wavelength. All the parts of the experiment were classical except for the lattice constant. All the parts came out of the shop. The atom velocity was specified with pulsed slotted discs. Hitler is to blame that we could not end these experiments in Hamburg. It was on the list of projects to be done.*" Here he implies that he had hoped to manufacture a regularly grooved grating in the shop as attempted in UZM, No. 11. Otto Stern would be happy to know that this experiment has recently been carried out in Berlin and is described in one of the following chapters by Wieland Schöllkopf.

In addition to the first observation of diffraction of massive particles and the confirmation of de Broglie's wave-particle duality, Otto Stern and his group made another important discovery. Already in the course of the first experiments, which led to clear diffraction peaks, Estermann and Stern in UzM No. 5 observed a series of four totally unexpected fairly sharp small minima upon azimuthal rotation of the target crystal [34]. They first considered that these could come from diffraction from unexpected structural modifications of the crystal. It was also speculated that they might be related to a layer of adsorbed molecules. To further investigate these anomalies, Frisch and Stern in UzM No. 23 constructed two new dedicated apparatus with additional degrees of freedom for moving the detector with respect to the crystal, but without velocity selection [35]. In one arrangement the diffraction angles could now be scanned out of the plane defined by the incoming beam and the crystal normal. In the other in-plane arrangement the incident angle could be varied and, in addition, the crystal could be rotated. In both arrangements, the distinct minima were confirmed and found to be even sharper. These they studied for He scattered from

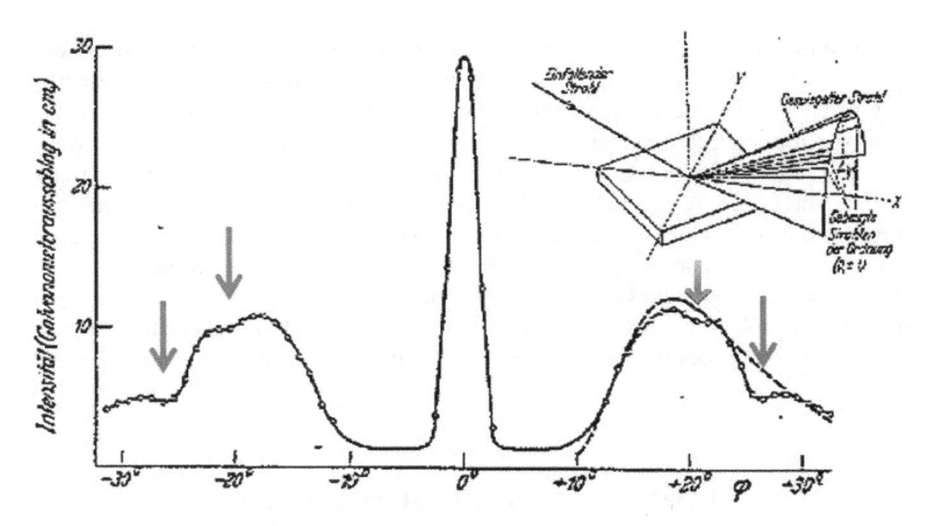

Fig. 9 Anomalous dips in the diffraction pattern, observed by Frisch in 1933 [35]. Three years later Lennard-Jones and Devonshire showed that the dips were due to atoms temporarily bound on the surface in a process called Selective Adsorption [36]

LiF over a wide range of angles in both arrangements. Minima were also found for H_2 on LiF and also for He scattered from NaF (Fig. 9). In summarizing their 1933 experiments in UzM No. 23 they write that they could not come up with a completely satisfactory explanation of these minima. They also report that the process depends in a specific way on the incident direction and energy of the particles. In 1933 in UzM No. 25, one of the last articles from the Hamburg group, Frisch was the only author since Stern, when the paper was submitted, left Germany three days later and had been busy with preparing his exodus. In this article Frisch further summarizes and characterizes the experimental conditions under which the minima appeared [38]. Here he correctly (see below) concludes that the adsorbed atoms are trapped in the two dimensional periodic force field of the surface from which they are diffusely reemitted. Since the perpendicular component of the motion of the atoms bound to the surface in the potential well must be quantized the incident atoms must initially have a specific incident direction and energy in order to be trapped.

Three years later in 1936 the careful and complete documentation of the experimental conditions under which the minima occurred enabled Lennard-Jones and Devonshire to develop the correct theory. The special angles and velocities at which the anomalies occurred were explained by the conditions of resonant trapping of the atoms by diffraction into the bound states of the atom-surface potential [36]. According to their theory: "*Atoms moving along the surface with the right energy and in the right direction may be diffracted as to leave the surface with positive energy and thus be evaporated. This is a new mechanism of evaporation which has not been previously expected.*" Since the minima only occur at special angles they named the new phenomenon Selective Adsorption (SA).

Since their discovery in 1933 and the explanation by Lennard-Jones and Devonshire, selective adsorption resonances have been extensively studied [39]. Presently, they provide the most sensitive and most direct probe of the bound states of the atom-surface potential and thereby the best method for determining the atom-surface potential. Since SA involves diffraction, the surface must have a sufficient corrugation for the effect to occur. Results are presently available for a wide variety of corrugated insulator surfaces and also for metal surfaces with sufficient corrugation as in the case of higher-index stepped surfaces [40]. Since 1981 several new types of resonances involving resonant *inelastic* processes, in which the bound states play an important role, have been found [39]. After a discussion of the elastic selective adsorption resonances the different types of inelastic resonances will be discussed below.

5 The Present Day Legacy of Otto Stern's Surface Scattering Experiments

On June 23, 1933 Stern and his colleagues Estermann, Frisch and Schnurmann were notified that they were discharged from the University, only Knauer, who was not Jewish, could remain. Stern was fortunate to soon after obtain a research professorship at Carnegie Institute of Technology in Pittsburgh. There he continued his molecular beam experiments, but did not continue the He atom surface diffraction experiments.[4] It appears that at the time he did not call attention or perhaps did not even realize the potential use of He atom diffraction for studying the structures of surfaces, which were largely unknown at the time. This we find surprising since in 1931 Thomas Johnson, who had reported diffraction of hydrogen atoms from LiF [42], wrote: "These experiments (diffraction of H_1, H_2 and He from LiF) are of interest not only because of their confirmation of the predictions of quantum mechanics, but also because they introduce the possibility of applying atom diffraction to investigations of the atomic constitution of surfaces. A beam of atomic hydrogen, ... has a range of wavelengths of the right magnitude ..., centering around 1 Å, and the complete absence of penetration of these waves will insure that the effects observed arise entirely from the outermost atomic layer".

Similarly, the significance and potential for surface physics of the 1927 electron scattering experiments of Davisson and Germer were not immediately realized. The experiments were only continued by Germer, who in 1929 reported on electron scattering to study gas adsorption [43]. Otherwise there were few immediate followers. One reason was that the apparatus used by Davisson and Germer were very fragile and prone to breakage. Also the preparation of metal surfaces was in the 1930s not well understood. Moreover, the depression in the 1930s and World War II halted

[4] In 1941, Bessey from the Carnegie Institute published a short note describing He and H_2 diffraction from LiF in which he thanks Otto Stern for suggesting the problem [41].

much of the fundamental research activities. One of the first to continue the experiments of Davisson and Germer was Harrison E. Farnsworth who was probably the first to use electron scattering to study the atomic structures of metal surfaces [44].

The remarkable experimental expertise of Otto Stern is well highlighted by the long time it took for others to repeat his He atom diffraction scattering experiments of 1929. The first post World War II attempt to repeat Stern's diffraction experiments was reported by J. Crews in 1962 [45] more than 30 years later. His angular distributions were not nearly as clearly resolved as in Stern's diffraction peaks. Also the 1968–1970 experiments of Okeefe et al [46, 47]. did not match up with the 1929 experiments. It was only after supersonic free jet expansion sources were introduced, with their inherent sharp velocity distributions, compared to the effusive atomic beam sources used by Stern, were comparable results achieved in 1973 [5]. Thus it took 44 years to arrive at a comparable technological-experimental level as achieved in the Hamburg group. This is even more surprising when it is realized that following the 1957 Sputnik shock the US embarked on a large program to compete scientifically with the Soviet Union. An important part of this program was to develop molecular beam research.

The outstanding experimental genius and foresight of Otto Stern was early on appreciated by the theoretician Max Born, who in 1919 had been Stern's colleague in Frankfurt. In his 1931 letter to the Nobel committee Born wrote: "According to my opinion Stern's achievement are far beyond those of other experimentalists through their conceptual boldness and also through the masterful overcoming of the experimental difficulties that I would like to propose no other physicist except him for the Nobel Prize."

The early history of atom and molecule surface scattering and diffraction experiments has been reviewed in the books by von Laue [48] and later in 1955 by Smith [49], and the more recent article by Comsa [50]. The first experimental studies of energy transfer in scattering from surfaces are described in the reviews by Beder [51], Stickney [52], and the author [53] and in 2018 in the monograph by Benedek and Toennies [39]. In 1969 Cabrera, Celli and Manson pointed out the possibility to observe single phonon excitations and their dispersion by inelastic diffraction of a He atom beam [53]. This stimulated several groups to carry out the corresponding scattering experiments. The first experiments to investigate the surface phonons of a crystal surface were performed by Brian Williams in Ottawa in 1971 [54]. He used essentially the same type of apparatus consisting of two diffraction surfaces that had been used by Estermann, Frisch and Stern for velocity selected diffraction in UzM No. 18. In the apparatus of Williams diffraction from the first crystal was used to select the velocity of the beam incident on the second crystal. The diffraction from the second crystal served to detect the inelastic change in velocity. Through an ingenious use of special kinematic conditions at certain incident and/or final directions additional small peaks in the angular distributions gave the first information on surface phonons on LiF(001) [54, 55]. Soon after in 1974 Boato and Cantini [56] also used high resolution angular distributions to study surface phonons with helium and neon scattered from LiF(001). Several groups also used time-of-flight inelastic scattering in further attempts to investigate the phonons on the surfaces of LiF [57] and on metals [58].

The first He atom inelastic time-of-flight experiments which were able to fully resolve single phonons and allowed the measurement of the dispersion curves of surface phonons out to the zone boundary were reported in 1981 by Brusdelylins, Doak and Toennies for the LiF surface [59] and in 1983 for a metal surface [60]. Similar measurements on metals using inelastic electron energy loss (EELS) were successfully carried out at about the same time [61, 62]. The He atom experiments were facilitated by the 1977 discovery that at high expansion pressures of about 100 bar free jet expansions of He have an inherent sharp velocity distribution corresponding to $\Delta \upsilon/\upsilon \approx 10^{-2}$[63]. It was no longer necessary to use diffraction and rotating discs to select the beam velocity.

The above experiments, continuing on the footsteps of Otto Stern, have established He atom scattering as a unique and indispensable surface science tool. A beam of He atoms in the thermal energy range (5–100 meV) is completely non-penetrating, chemically inert and produces no mechanical damage. Because of its matter-wave property it is the ideal method to project out of the chaotic vibrating surface by inelastic diffraction the phonon dispersion curves. Electron beams have also this property but since the electrons at the same wave length have much higher energies the electrons penetrate the surface and can damage the crystal. The modern experiments in which He atom diffraction is used to study the surface structures of insulating, semiconductor and metal surfaces have been reviewed by Rieder and Engel [64] and by Farias and Rieder [65]. The Helium Atom Scattering (HAS) studies of the phonon dispersion curves of clean surfaces and the vibrations of adsorbate-covered surfaces have very recently been surveyed in the 2018 book by Giorgio Benedek and J. P. Toennies entitled "Atomic Scale Dynamics at Surfaces: Theory and Experimental Studies with Helium Atom Scattering" [39]. There also the very recent understanding that He atoms interact with the electron densities at the surface and provide detailed information on the electron-phonon coupling constant is discussed.

6 New Applications of Matter-Wave Diffraction from Manufactured Nanoscopic Gratings

The advent of nanotechnology in the last 30–40 years has opened up new opportunities to utilize the wave-particle duality for investigations of the physical properties of atoms, molecules and clusters. Some very recent developments in this area are covered in the review articles in this book following this introductory historical review. Here only some early pioneering experiments are dealt with.

Free standing transmission gratings with a slit spacing commensurate with wave lengths in the soft X-ray and the extreme ultraviolet regime, where traditional optical elements are opaque, were first developed for spectroscopy. With decreasing structural dimensions, nanostructured transmission gratings have made it possible to carry over to massive particles with much smaller wave lengths the interference phenomena which had been exploited for light.

Keith, Schattenburg, Smith and Pritchard at MIT were the first in 1988 to report the diffraction of atoms from a fabricated transmission grating [66]. In their experiment Na atoms from an in Ar seeded beam with a de Broglie wave length of 0.017 nm (0.17 Å) were diffracted by an angle of about 70 μrad with an angular resolution of 25 μrad(!) after passing through a specially fabricated 0.2 μm period gold grating with slits and bars of 0.1 μm width. Previously, the same group had also shown that atoms could be diffracted from a standing wave of near resonant light [67]. Subsequently, in 1991, Carnal and Mlynek demonstrated a simple interferometer based on Young's double slit experiment using 2 μm wide slits [68]. Since these initial experiments transmission grating diffraction has been reported for molecules such as Na_2 [69], C_{60} [70, 71], $C_{60}F_{48}$ and $C_{44}H_{30}N_4$ and also for small helium clusters [72–74], with up to about 50 atoms [74–76].

The author's group in 1994 applied transmission grating diffraction as a type of mass spectrometer to establish the existence of the He atom dimer [72, 73]. The existence of the He dimer was long questioned since the long range attractive potential could not be predicted with sufficient accuracy to establish if the very weak attraction between the atoms would be strong enough to support a bound state. A 1993 claim based on mass spectrometer detection [77] was subsequently questioned [78, 79]. Figure 10 shows the diffraction apparatus used by our group [80]. The mass selection comes about since the de Broglie wave length and the diffraction angle are inversely proportional to the mass of the diffracted particles. Since only wavelets which pass through the slits coherently without any interaction with the grating bars can contribute to the diffraction peaks this type of mass spectrometer is completely non-destructive.

Figure 11 displays some diffraction patterns taken at different source temperatures and a high resolution measurement showing well resolved dimer and trimer

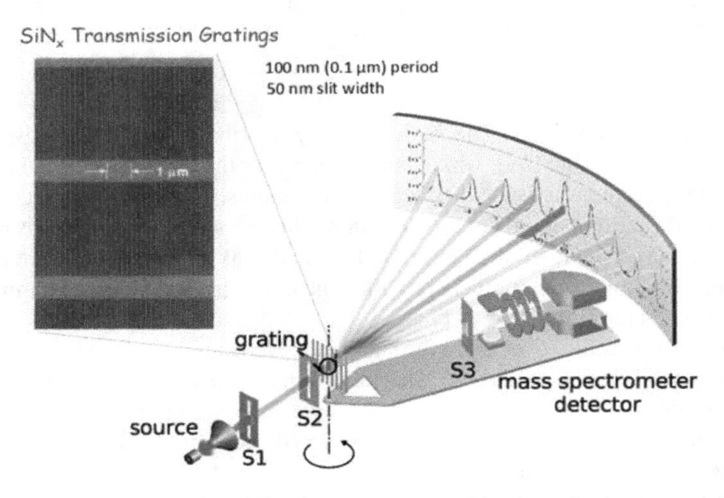

Fig. 10 The transmission grating diffraction apparatus used by the author's group to detect the helium dimer and other small He clusters [80]

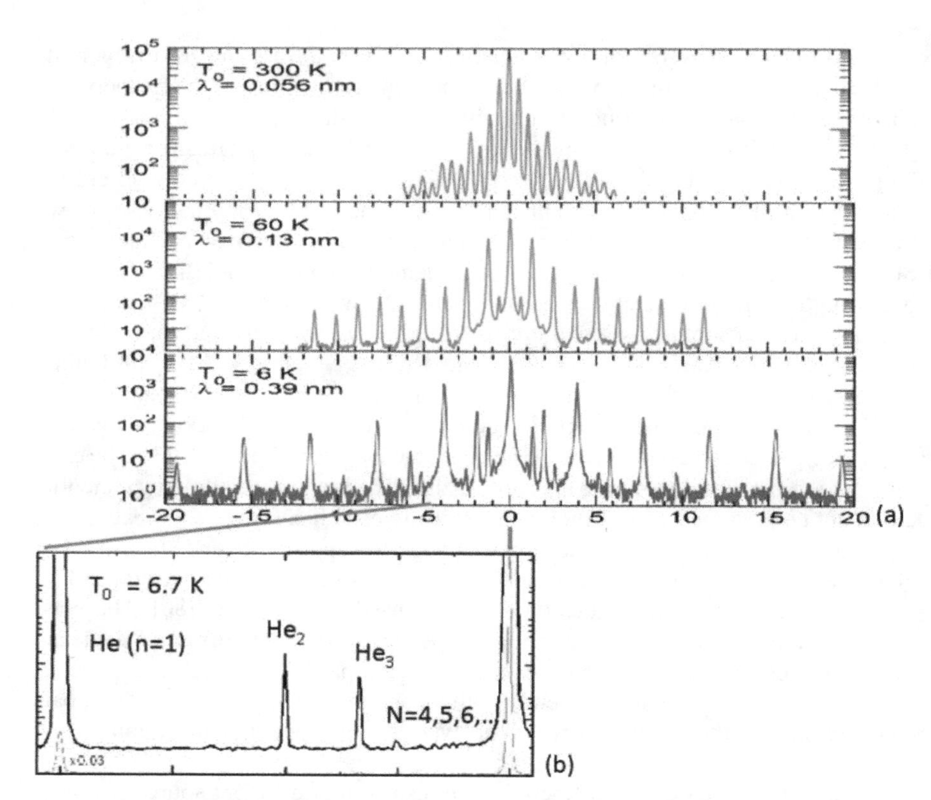

Fig. 11 **a** He atom diffraction patterns measured at three different decreasing source temperatures with increasingly large de Broglie wavelengths. At the lowest temperature between the central specular peak and the first order atom diffraction peaks at -4 and +4 degrees additional diffraction peaks appear which correspond to the dimer, trimer and tetramer of helium. **b** The same diffraction peaks measured with a much increased angular resolution [81]

first order diffraction peaks. The sharp velocity distributions of high-pressure free jet expansions of helium [63], which is also found for the clusters of helium, greatly facilitated the resolution of the diffraction experiments. The partly resolved anomalies in the diffraction patterns of larger helium clusters with up to 50 atoms revealed unexpected maxima and minima instead of a broad peak expected for liquid clusters. The corresponding magic numbers provided the first evidence for the quantum levels of these small superfluid clusters [74].

It was even possible to measure the size of the He-dimer from the intensity distribution of the diffraction peaks [82]. The weak bond of the dimer makes it sensitive to even the slightest interaction with the grating bars. Depending on the size of the dimer the effective slit and coherence width are reduced accordingly. The effective width was experimentally determined from the *slit function* which is the envelope over all the diffraction peaks out to high order. From extensive measurements of the diffraction patterns of the He atom and He dimer it was found that the dimer diffraction

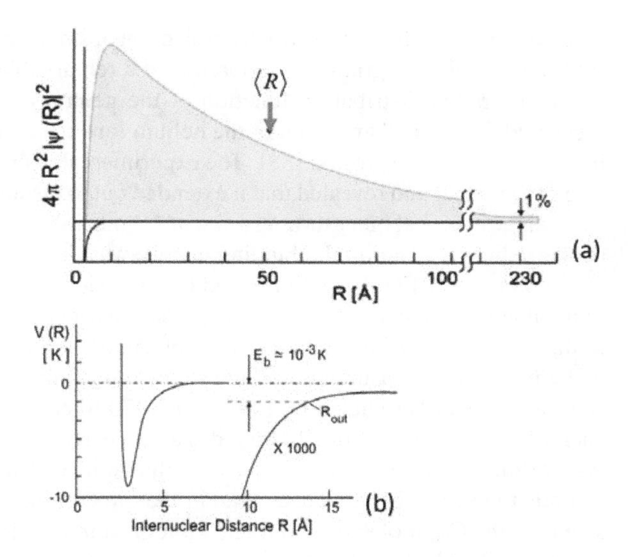

Fig. 12 **a** The probability amplitude of the He dimer which is compatible with the measured mean radius of the dimer of 5.2 ± 0.4 nm [82]. **b** An expanded view of the potential showing that the highly quantum dimer tunnels far beyond the classical outer turning point located at about 1.4 nm

peaks were associated with a significantly smaller slit width, which corresponds to a larger size, than that of the atoms. With the aid of a many-body quantum scattering theory the difference in slit width could be referred to the mean internuclear distance of the dimer $\langle R \rangle$ [83]. The extreme sensitivity is illustrated by the fact that $\langle R \rangle$ was found to be 5.2 ± 0.4 nm. The uncertainty was only 0.4% when compared to the 100 nm overall slit width of the grating.[5] This experiment established that the dimer is the largest of all ground state molecules and is about 70 times larger than the H_2 molecule (Fig. 12). It also provided the first measurement of the dimer van der Waals bond which is still an important benchmark for quantum chemists. From the bond distance the binding energy was calculated to be only $96^{+26}_{-17} \times 10^{-9}$ eV($1.1^{+0.3}_{-0.2}$ mK) [82] which corresponds to a scattering length of about s = 100 Å.

In the case of other atoms and tightly bound molecules the effective slit and coherence width depend on the long range van der Waals interaction with the solid surface given by $V(l) = -C_3/l^3$, where l is the distance of the particle from the slit surface as it passes through the slits of the grating. Both the magnitude of the corresponding reduction in the effective slit width and its velocity dependence were fitted with a scattering theory to provide values of the van der Waals constant C_3 [84]. The values obtained showed the expected linear increase of C_3 with the polarizability of the particle increasing in the order of He, Ne, D_2, Ar und Kr. These experiments represent the first quantitative measurements of C_3. The method was subsequently used to determine the C_3 constants of the alkali atoms [85, 86] and of metastable atoms [87] and to set limits on the strength of the non-Newtonian gravity at short length scales [86].

[5] The method is so sensitive that the effect of a monolayer of adsorbed Xe in narrowing the slit width by about 7 Å could be measured by comparing the slit width with a cold and a hot grating in the presence of ambient Xe[(unpublished].

Quite recently, the non-destructive selection of small helium clusters by diffraction from a transmission gratings has facilitated a remarkable study. In this experiment the actual radial distribution function of the neutral atoms in the dimer could be measured from the distribution of the helium ions released in the femtosecond laser induced Coulomb explosion [88]. The experimental distribution confirmed the large size of the dimer and revealed that it extended out to distances of more than 23 nm far beyond the classical outer turning point at 1.4 nm [88]. From the exponential fall-off of the radial distribution the binding energy was determined to be $151.9 \pm 13.3 \times 10^{-9}$ eV(1.5 ± 0.13 mK) with the highest precision so far. With the same apparatus in another remarkable experiment the radial distributions of the three helium atoms in the first excited Efimov state of helium trimer could also be measured [89]. This is the first direct measurement of the size of an Efimov state. This unique quantum state was already predicted by Efimov in 1970 to occur for three bosons, of which each of the pairs are critically bound with a nearly vanishing binding energy [90]. As a result of the smaller binding energy the radial distribution of the Efimov trimer extends to even larger distances than in the dimer making the trimer even larger in size than the C_{60} molecule and many biological molecules and even viruses.

The de Broglie wavelength of atoms has also been the basis of using transmission Fresnel zone plates for focusing atomic beams. The focusing via Fresnel zone plates with many concentric slits was first demonstrated by Carnel et al. already in 1991 [68]. Using easily detected electronically excited He atoms they were able to focus down to a spot size of 15–20 μm with an intensity of 0.5 counts/s. In 1999, Doak et al. from our laboratory using a miniature 0.27 mm overall dia. zone plate with a smallest outermost slit width of 100 nm (this is also the predicted diffraction limited resolution) achieved with neutral He atoms a 2 μm dia spot with an intensity of 350 counts/s [91]. Recently, the spot size was reduced down to 1 μm but with an intensity of only 125 counts/s [92].

The first interferometers based on matter wave diffraction were with electrons [93–95] and neutrons [96, 97]. In both experiments the passage through a well-ordered crystal was used as the diffracting element. A Mach-Zehnder interferometer using transmission gratings with Na atoms was first reported by Keith et al. [98]. Subsequently, the same interferometer was used to demonstrate the loss of coherence by scattering photons from one of the paths through the interferometer [67]. Since these seminal demonstration experiments interferometry with atoms has become a wide field of activity stimulated by advances in laser and nanotechnology and laser cooling techniques [99, 100]. The recent advances are the subject of the article by Stefan Gerlich and colleagues in the next chapter.

The construction of a simple robust and compact three grating Mach-Zehnder interferometer developed in our group is shown in Fig. 13 [101]. A homogeneous electric field in one of the interfering arms was used to shift the phase in one of the two separate branches with respect to the other branch. Figure 14 shows the resulting interference fringes. In this experiment a novel extremely accurate lower limit on the velocity half-width of a He atom free jet beam could be demonstrated. In one branch of the interferometer an electric field was applied. With increasing field strength one half of the beam was shifted in phase with respect to the beam without an applied

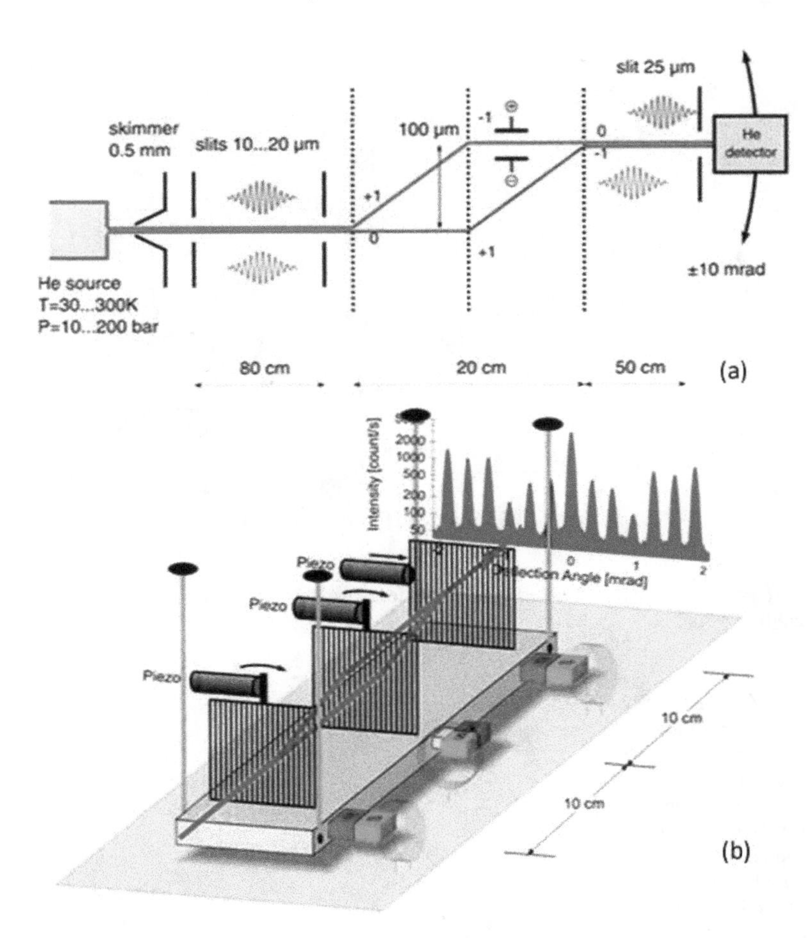

Fig. 13 **a** Schematic diagram of the compact Mach-Zehnder Interferometer constructed by the Göttingen group. **b** Perspective view of the interferometer showing the piezos to adjust the gratings and the magnetic vibration stabilization to reduce the vibrations to less than 1×10^{-6} m [101]

field. The interference pattern of Fig. 14 shows the interferences as one half of the cold 6 K beam with a de Broglie wave length of 0.4 nm (4 Å) had been shifted with respect to the other half by 25 wave lengths. The largest shift corresponds to a lower limit on the parallel coherence length of 25×4 Å $= 100$ Å, which corresponds to a velocity half width of only 10 m/sec.

A Mach-Zehnder interferometer also provides a precise method for measuring polarizabilities as was demonstrated for the highly polarizable sodium [102] and lithium [103]. Figure 15 shows the results of an interferometer experiment designed to measure the polarizability of the weakly bound He dimer. In the diffraction pattern at the outlet of the interferometer diffraction peaks due to the dimer could be identified between the more intense peaks of the He atom component. From the phase

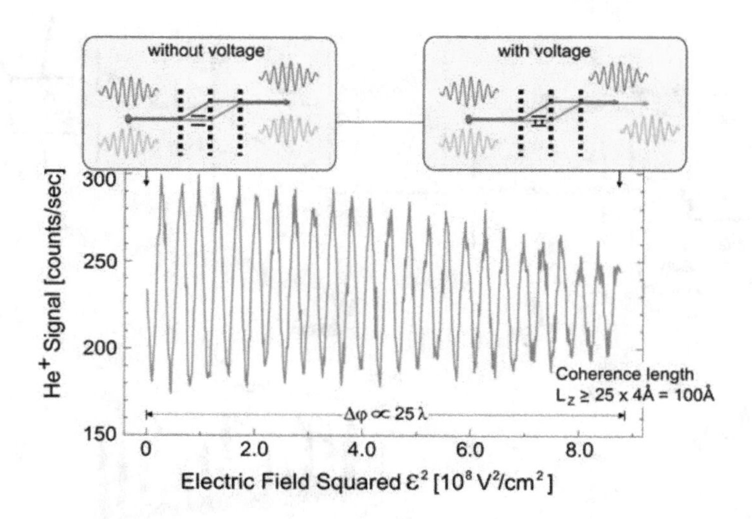

Fig. 14 Interference fringes obtained with a free jet expanded He atom beam with a de Broglie wavelength of 4 Å [101]

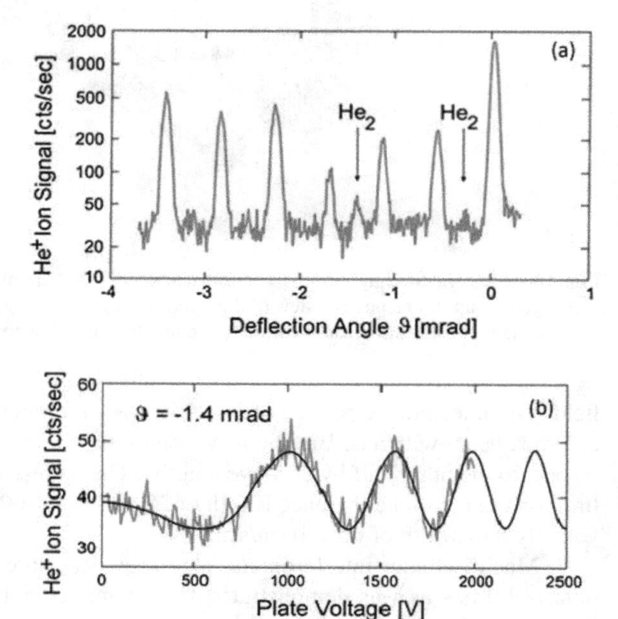

Fig. 15 The polarizability of the He dimer is measured in the same arrangement as Figs. 13 and 14 [101]. **a** From the diffraction pattern at the output of the interferometer, a small dimer peak could be identified at -1.4 mrad. **b** From the interference fringes measured at this angle with increasing voltage the polarizability of the He dimer was found to be somewhat smaller than that of two separate atoms

shift with voltage the polarizability of both the He atom and the dimer could be measured relative to each other. Since the atom polarizability is well-known from both metrological experiments and from theory the polarizability of the He dimer could be calibrated by comparison with the He atom beam component and found to be

$$\propto (He_2) = 0.30884 \pm 0.001883\,\text{Å}^3,$$

which is to be compared with the polarizability of two He atoms:

$$\propto (2He) = 0.3113 \pm 0.0023\,\text{Å}^3.$$

The smaller polarizability of the dimer is expected since the electrons in the dimer are very slightly more bound than in the fully separated atoms. This preliminary result serves to illustrate the remarkable sensitivity of the method. In a similar manner the comparatively large polarizabilities have been measured of the alkali atoms Li [104] and Na, K, and Rb [105].

7 Summary

In the 90 years since Otto Stern had demonstrated that wave-particle duality also applied to massive particles, the wave nature of atoms and molecules has found widespread applications in both physics and chemistry. The first experiments of He atom diffraction scattering from LiF crystals by Otto Stern in 1929 have since evolved into a gentle nondestructive and universal tool for the determination of the structure of clean and adsorbate-covered surfaces and for the measurement of the phonon dispersion curves at the surfaces of a wide range of different types of solids [39]. In the latter application, He atom scattering is the ideal complement to the scattering of neutrons, which since they pass through the crystal virtually unhindered, are only sensitive to the bulk phonons.

Since Otto Stern's days new nanotechnology advances have opened up an entire new field of experiments based on matter wave behavior of atoms and molecules. As discussed here these encompass on the one hand the non-destructive mass analysis of fragile clusters and on the other hand the precision interferometry of atomic and molecular properties. Certainly, Stern would have been happy to learn about the many wonderful and important applications of matter waves of atoms and molecules discussed here and described in the following chapters.

In this connection, the author, after having been occupied with the impact of Otto Stern's 1929 diffraction experiments, often wonders whether Stern at the time realized the many future important applications arising out of his pioneering experiments. It is interesting that he had attempted and had hoped to carry out diffraction experiments from man-made ruled gratings, only recently realized and described by Wieland Schöllkopf in the following chapter.

Otto Stern was definitely very much aware of the fundamental importance of his experiments in relation to the electron diffraction experiments and the implications for quantum theory as expressed in his 1943 Nobel lecture: "*With respect to the differences between the experiments with electrons and molecular rays, one can say that the molecular ray experiments go farther. Also the mass of the moving particle is varied (He, H_2). But the main point is again that we work in such a direct primitive manner with neutral particles. These experiments demonstrate clearly and directly the fundamental fact of the dual nature of rays of matter. It is no accident that in the development of the theory the molecular ray experiments played an important role. Not only the actual experiments were used, but also molecular ray experiments carried out only in thought. Bohr, Heisenberg, and Pauli used them in making clear their points on this direct simple example of an experiment...*" Here Otto Stern calls attention to the important impact of de Broglie's theory and his experiments on the development of quantum theory especially by Erwin Schrödinger. To do justice this aspect would be beyond the scope of this article.

Acknowledgements The author would like to thank Horst Schmidt-Böcking and Bretislav Friedrich for stimulating my interest in the life and scientific work of Otto Stern. I wish to also thank Katrin Glormann for her great care and enduring patience in preparing the manuscript and Bretslav Friedrich for a careful reading and corrections of the final manuscript.

References

1. O. Stern, Z. Physik **39**, 751 (1926)
2. L. de Broglie, *Thèse de doctoral,* Mason, Paris (1924); reprinted 1963; also Annales de Physique**3**, 22 (1925)
3. O. Stern, Naturwissenschaften **17**, 391 (1929)
4. C. Davisson, L.H. Germer, Nature **119**, 558 (1927)
5. G. Boato, P. Cantini, U. Garibaldi, A.C. Levi, L. Mattera, R. Spadacini, G.E. Tommei, J. Phys. C Solid State **6**, L394 (1973)
6. L. de Broglie, J. de Physique **3**, 422 (1922)
7. L. de Broglie, Comptes Rendus **175**, 881 (1922)
8. L. de Broglie, Comptes Rendus **177**, 507 (1923)
9. L. de Broglie, Nature **112**, 540 (1923)
10. L. de Broglie, Philos. Mag. **47**, 446 (1924)
11. L. de Broglie, Ann. Phys. **3**, 22 (1925)
12. H.A. Medicus, Phys. Today **27**, 38 (1974)
13. E. MacKinnon, Ann. J. Phys. **44**, 1047 (1976)
14. A. Einstein, Preuss. Akad. Wiss. Mathem.-Naturwisse. Kl. **23**, 3 (1925)
15. C. Davisson, C.H. Kunsman, Science **54**, 522 (1921)
16. E. Davisson, C.H. Kunsman, Phys. Rev. **22**, 242 (1923)
17. W. Elsassser, Naturwissenshaften 13, 711 (1925)
18. H. Rubin, *A Biographical Memoir of Walter M. Elsasser 1904–1991* (National Academy of Sciences, Washington D.C., 1995)
19. F. Hund, *Geschichte der Quantentheorie,* Bibliographisches Institut Mannheim (1967)
20. M. Born, Z. Phys. **38**, 863 (1926)
21. M. Born, *Experiments and Theory in Physics* (Dover, 1943)
22. R.K. Gehrenbeck, Phys. Today (January 1978), 34

23. R. Schlegel, R.K. Gehrenbeck, Phys. Today **31**, 9 (1978)
24. C. Davisson, L.H. Germer, Phys. Rev. A **30**, 705 (1927)
25. G.P. Thomson, A. Reid, Nature **119**, 890 (1927)
26. G.P. Thomson, A. Reid, Proc. Roy. Soc. (London) A 117 (1928) 601
27. S. Weinberg, *The Trouble with Quantum Mechanics,* The New York Times Review of Books (January 19, 2017)
28. J.P. Toennies, H. Schmidt-Bocking, B. Friedrich, J.C.A. Lower, Ann. Phys. **523**, 1045 (2011)
29. O. Stern, Phys. Z. **21**, 582 (1920)
30. W. Gerlach, O. Stern, Z. Phys. **9**, 349 (1922)
31. F. Knauer, O. Stern, Z. Phys. **53**, 779 (1929)
32. T.H. Johnson, J. Franklin Inst. **206**, 301 (1928)
33. A. Ellet, H.F. Olsen, Phys. Rev. **31**, 643 (1928)
34. I. Estermann, O. Stern, Z. Phys. **61**, 95 (1930)
35. R.O. Frisch, O. Stern, Z. Phys. **84**, 430 (1933)
36. J.E. Lennard-Jones, A.F. Devonshire, Nature **137**, 1069 (1936)
37. R. Jost, ETH-Bibliothek Zürich, Archive (1961)
38. R.O. Frisch, Z. Phys. **84**, 443 (1933)
39. G. Benedek, J.P. Toennies, *Atomic Scale Dynamics at Surfaces: Theory and Experimental Studies with Helium Atom Scattering* (Springer, 2018)
40. G. Vidali, G. Ihm, H.Y. Kim, M.W. Cole, Surf. Sci. Rep. **12**, 133 (1991)
41. W.H. Bessey, Phys. Rev. **59**, 459 (1941)
42. T.H. Johnson, Phys. Rev. **37**, 847 (1931)
43. L.H. Germer, Z. Phys. **54**, 408 (1929)
44. H.E. Farnsworth, Phys. Rev. **34**, 679 (1929)
45. J.C. Crews, J. Chem. Phys. **37**, 2004 (1962)
46. D.R. Okeefe, J.N. Smith, R.L. Palmer, H. Saltsburg, J. Chem. Phys. **52**, 4447 (1970)
47. R. Okeefe, R.L. Palmer, H. Saltsburg, J.N. Smith, J. Chem. Phys. **49**, 5194 (1968)
48. M. von Laue, *Materiewellen und ihr Interferenzen* (Akadem. Verl.-Ges. Becker & Erler, Geest und Portig, 1944)
49. K.F. Smith, *Molecular Beams* (Wiley, 1955)
50. G. Comsa, Surf. Sci. **299**, 77 (1994)
51. E.C. Beder, Advan. At. Mol. Phys. **3**, 205 (1968)
52. R.E. Stickney, Advan. At. Mol. Phys. **3**, 143 (1968)
53. N. Cabrera, V. Celli, R. Manson, Phys. Rev. Lett. **22**, 346 (1969)
54. B.R. Williams, J. Chem. Phys. **55**, 3220 (1971)
55. B.F. Mason, B.R. Williams, J. Chem. Phys. **56**, 1895 (1972)
56. G. Boato, P. Cantini, in *Dynamics Aspects of Surface Physics*, ed. by F.O. Goodman (1974), p. 707
57. S.S. Fisher, J.R. Bledsoe, J. Vac. Sci. Technol. **9**, 814 (1972)
58. S.C. Yerkes, D.R. Miller, J. Vac. Sci. Technol. **17**, 126 (1980)
59. G. Brusdeylins, R.B. Doak, J.P. Toennies, Phys. Rev. Lett. **46**, 437 (1981)
60. R.B. Doak, U. Harten, J.P. Toennies, Phys. Rev. Lett. **51**, 578 (1983)
61. S. Lehwald, J.M. Szeftel, H. Ibach, T.S. Rahman, D.L. Mills, Phys. Rev. Lett. **50**, 518 (1983)
62. J.M. Szeftel, S. Lehwald, H. Ibach, T.S. Rahman, J.E. Black, D.L. Mills, Phys. Rev. Lett. **51**, 268 (1983)
63. J.P. Toennies, K. Winkelmann, J. Chem. Phys. **66**, 3965 (1977)
64. T. Engel, K.-H. Rieder, *Structural Studies of Surfaces* (Springer-Verlag, Berlin-Heidelberg-New York, 1982)
65. D. Farias, K.H. Rieder, Rep. Prog. Phys. **61**, 1575 (1998)
66. D.W. Keith, M.L. Schattenburg, H.I. Smith, D.E. Pritchard, Phys. Rev. Lett. **61**, 1580 (1988)
67. M.S. Chapman, T.D. Hammond, A. Lenef, J. Schmiedmayer, R.A. Rubenstein, E. Smith, D.E. Pritchard, Phys. Rev. Lett. **75**, 3783 (1995)
68. O. Carnal, J. Mlynek, Phys. Rev. Lett. **66**, 2689 (1991)
69. M.S. Chapman, C.R. Ekstrom, T.D. Hammond, J. Schmiedmayer, S. Wehinger, D.E. Pritchard, Phys. Rev. Lett. **74**, 4783 (1995)

70. M. Arndt, O. Nairz, J. Vos-Andreae, C. Keller, G. van der Zouw, A. Zeilinger, Nature **401**, 680 (1999)
71. Y.Y. Fein, P. Geyer, F. Kialka, S. Gerlich, M. Arndt, Phys. Rev. Res. **1**, 033158 (2019)
72. W. Schöllkopf, J.P. Toennies, Science **266**, 1345 (1994)
73. W. Schollkopf, J.P. Toennies, J. Chem. Phys. **104**, 1155 (1996)
74. R. Guardiola, O. Kornilov, J. Navarro, J.P. Toennies, J. Chem. Phys. **124**, 084307 (2006)
75. R. Brühl, R. Guardiola, A. Kalinin, O. Kornilov, J. Navarro, T. Savas, J.P. Toennies, Phys. Rev. Lett. **92**, 185301 (2004)
76. L. Hackermuller, S. Uttenthaler, K. Hornberger, E. Reiger, B. Brezger, A. Zeilinger, M. Arndt, Phys. Rev. Lett. **91** (2003)
77. F. Luo, G.C. Mcbane, G.S. Kim, C.F. Giese, W.R. Gentry, J. Chem. Phys. **98**, 3564 (1993)
78. E.S. Meyer, J.C. Mester, I.F. Silvera, J. Chem. Phys. **100** (1994)
79. F. Luo, G.C. Mcbane, G. Kim, C.F. Giese, W.R. Gentry, J. Chem. Phys. **100**, 4023 (1994)
80. O. Kornilov, J.P. Toennies, Europhys. News **38**, 22 (2007)
81. J.P. Toennies, Mol. Phys. **111**, 1879 (2013)
82. R.E. Grisenti, W. Schöllkopf, J.P. Toennies, G.C. Hegerfeldt, T. Köhler, M. Stoll, Phys. Rev. Lett. **85**, 2284 (2000)
83. G.C. Hegerfeldt, T. Kohler, Phys. Rev. A **57**, 2021 (1998)
84. R.E. Grisenti, W. Schöllkopf, J.P. Toennies, C.C. Hegerfeldt, T. Köhler, Phys. Rev. Lett. **83**, 1755 (1999)
85. V.P.A. Lonij, C.E. Klauss, W.F. Holmgren, A.D. Cronin, J. Phys. Chem. A **115**, 7134 (2011)
86. S. Lepoutre, V.P.A. Lonij, H. Jelassi, G. Trenec, M. Buchner, A.D. Cronin, J. Vigue, Eur. Phys. J. D **62**, 309 (2011)
87. R. Brühl, P. Fouquet, R.E. Grisenti, J.P. Toennies, G.C. Hegerfeldt, T. Köhler, M. Stoll, C. Walter, Europhys. Lett. **59**, 357 (2002)
88. S. Zeller, M. Kunitski, J. Voigtsberger, A. Kalinin, A. Schottelius, C. Schober, M. Waitz, H. Sann, A. Hartung, T. Bauer, M. Pitzer, F. Trinter, C. Goihl, C. Janke, M. Richter, G. Kastirke, M. Weller, A. Czasch, M. Kitzler, M. Braune, R.E. Grisenti, W. Schöllkopf, L.P.H. Schmidt, M.S. Schöffler J.B. Williams, T. Jahnke, R. Dörner, P. Natl. Acad. Sci. USA **113**, 14651 (2016)
89. M. Kunitski, S. Zeller, J. Voigtsberger, A. Kalinin, L.P.H. Schmidt, M. Schöffler, A. Czasch, W. Schöllkopf, R.E. Grisenti, T. Jahnke, D. Blume, R. Dörner, Science **348**, 551 (2015)
90. V. Efimov, Sov J. Nucl. Phys. **12**, 1080 (1970)
91. R.B. Doak, R.E. Grisenti, S. Rehbein, G. Schmahl, J.P. Toennies, C. Wöll, Phys. Rev. Lett. **83**, 4229 (1999)
92. S.D. Eder, T. Reisinger, M.M. Greve, G. Bracco, B. Holst, New J. Phys. **14**, 073014 (2012)
93. L. Marton, J.A. Simpson, J.A. Suddeth, Phys. Rev. **90**, 490 (1953)
94. L. Marton, J.A. Simpson, J.A. Suddeth, Rev. Sci. Instrum. **25**, 1099 (1954)
95. G. Moellenstedt, H. Dueker, Die Naturwissenschaften **42**, 41 (1955)
96. H. Maier Leibnitz, T. Springer, Z. Phys. **167**, 386 (1962)
97. H. Rauch, A. Werner, *Neutron Interferometry: Lessons in Experimental Quantum Mechanics* (Clarendon Press, Oxford, 2000)
98. D.W. Keith, C.R. Ekstrom, Q.A. Turchette, D.E. Pritchard, Phys. Rev. Lett. **66**, 2693 (1991)
99. A.D. Cronin, J. Schmiedmayer, D.E. Pritchard, Rev. Mod. Phys. **81**, 1051 (2009)
100. K. Hornberger, S. Gerlich, P. Haslinger, S. Nimmrichter, M. Arndt, Rev. Mod. Phys. **84**, 157 (2012)
101. R. Brühl, R.B. Doak, J.P. Toennies, (unpublished)
102. C.R. Ekstrom, J. Schmiedmayer, M.S. Chapman, T.D. Hammond, D.E. Pritchard, Phys. Rev. A **51**, 3883 (1995)
103. A. Miffre, M. Jacquey, M. Buchner, G. Trenec, J. Vigue, Eur. Phys. J. D **38**, 353 (2006)
104. A. Miffre, M. Jacquey, M. Buchner, G. Trenec, J. Vigue, Phys. Rev. A **73**, 011603(R) (2006)
105. W.F. Holmgren, M.C. Revelle, V.P.A. Lonij, A.D. Cronin, Phys. Rev. A **81**, 053607 (2010)

3

When Liquid Rays Become Gas Rays: Can Evaporation ever be Non-Maxwellian?

Gilbert M. Nathanson

Abstract A rare mistake by Otto Stern led to a confusion between density and flux in his first measurement of a Maxwellian speed distribution. This error reveals the key role of speed itself in Stern's development of "the method of molecular rays". What if the gas-phase speed distributions are not Maxwellian to begin with? The molecular beam technique so beautifully advanced by Stern can also be used to explore the speed distribution of gases evaporating from liquid microjets, a tool developed by Manfred Faubel. We employ liquid water and alkane microjets containing dissolved helium atoms to monitor the speed of evaporating He atoms into vacuum. While most dissolved gases evaporate in Maxwellian speed distributions, the He evaporation flux is *super*-Maxwellian, with energies up to 70% higher than the flux-weighted average energy of $2\,RT_{\text{liq}}$. The explanation of this high-energy evaporation involves two beautiful concepts in physical chemistry: detailed balancing between He atom evaporation and condensation (starting with gas-surface collisions) and the potential of mean force on the He atom (starting with He atoms just below the surface). We hope that these measurements continue to fulfill Stern's dream of the "directness and simplicity of the molecular ray method."

1 Introduction: J. C. Maxwell and Otto Stern

Otto Stern's first publication, in 1920, described an ingenious Coriolis measurement of the root-mean-square (rms) speed of a Maxwellian distribution of silver atoms emitted from a hot oven ("gas rays") [1]. It is remarkable that this distribution had not been measured before, but even more remarkable was the correction to Stern's article later in 1920. Stern's postdoctoral advisor, Albert Einstein, pointed out to Stern that he had calculated the rms speed, $\langle c^2\rangle^{1/2}_{\text{density}} = (3RT/m)^{1/2}$ using the density weighting $n(c)$ instead of the flux (velocity) weighted average $\langle c^2\rangle^{1/2}_{\text{flux}} = (4RT/m)^{1/2}$, where the

G. M. Nathanson (✉)
Department of Chemistry, University of Wisconsin-Madison, 1101 University Avenue, Madison, WI 53706, USA
e-mail: nathanson@chem.wisc.edu

flux $J(c) = c \cdot n(c)$. Stern immediately published a correction that agreed more closely with his measured value [2], and 27 years later published a measurement of the full distribution [3]. It is heartening to know that even the great Otto Stern made mistakes, although it took someone of the stature of Einstein to correct him! (See Chap. 5 for more history.) In a sense, this chapter starts with Stern's mistake by exploring the nature of speed distributions, but with a focus on the speeds of evaporating gases dissolved in liquid microjets in vacuum ("liquid rays"). Our discussion of non-Maxwellian evaporation weaves a tale that involves two beautiful concepts in physical chemistry, namely detailed balancing between condensation and evaporation and the potential of mean force for a dissolved gas in solution.

The Maxwellian properties of number-density and flux distributions are thoroughly summarized by David and Comsa, a review article I highly recommend [4]. These two distributions can be imagined using the fingers on one hand. Cup the air within your fist: the molecules trapped inside have a Maxwellian speed distribution given by $n(c) \sim c^2 e^{-mc^2/2RT} n_{gas}$. Here $n(c)\mathrm{d}c$ is the number of molecules per unit volume in a narrow speed interval dc. Now make an "O" with your thumb and forefinger: the speed distribution of molecules passing through the "O" is instead the flux (speed-weighted) distribution, $J(c, \theta) \sim c^3 e^{-mc^2/2RT} \cos\theta \, n_{gas}$, where θ is the polar angle. In this case, $J(c, \theta)\sin\theta \, \mathrm{d}\theta \, \mathrm{d}\phi \, \mathrm{d}c$ is the number of molecules passing through a unit area per second per unit speed and solid angle interval. This distribution is shifted toward higher speeds (c^3 vs. c^2) because faster molecules traverse the area of the "O" more frequently than do slower molecules. $J(c, \theta)$ is also weighted by $\cos\theta$ because the normal velocity, $c_z = c \cdot \cos\theta$, is the component that transports the gas molecule to the surface formed by the "O", such that the integrated gas-surface collision frequency with a unit area is given by $(RT/2\pi m)^{1/2} n_{gas}$ [5]. Next situate your "O" over the surface of a glass of water: the flux of water vapor or other gas molecules striking the surface, as pictured as in Fig. 1, is just the same $J(c, \theta)$. This review addresses how the probabilities of dissolution and evaporation vary with the translational energy of the gas molecule, and what this dependence tells us about the mechanisms of solvation.

Maxwell's seminal 1860 article derived the number-density speed distribution of molecules that bears his name and often that of Boltzmann [6]. In a later 1879 article, Maxwell included comments on collisions of molecules with surfaces [7]. He categorized gas molecules striking a surface in two distinct ways: adsorption, which refers to the trapping of molecules at the surface (bound in a physisorption or chemisorption well), and reflection, which corresponds to an immediate, direct bounce from the surface. The fact that not all gases stick upon collision with a surface was in fact proved by Estermann and Stern in their celebrated study of the diffraction of helium atoms from the surface of crystalline lithium fluoride in 1930 [8]. Maxwell's and Stern's paths intersected more than once!

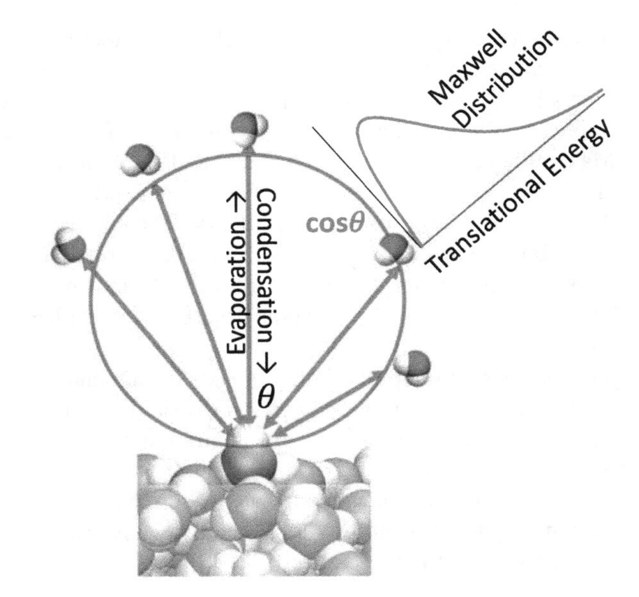

Fig. 1 Condensation and evaporation are reverse processes. Water molecules strike the surface in a cosine angular distribution and velocity(flux)-weighted Maxwellian distribution of translational energies. When every approaching water molecule sticks, the evaporation distribution is also cosine and Maxwellian. The simulation snapshot of the surface of water is adapted with permission from P. Jungwirth, Water's wafer-thin surface, Nature, **474**, 168–169 (2011)

2 Condensation and Evaporation as Reverse Processes

We now know that molecules colliding with a surface interact in numerous ways, as summarized in recommended reviews [9–18]. During a single or multi-bounce nonreactive collision, these pathways include not only translational energy exchange but also vibrational, rotational, and electronic transitions (including spin-orbit) in the gas-phase molecule and in the surface and subsurface molecules within the collision zone. The range of energy exchange can vary from zero (elastic collisions such as occurs in diffraction) through production of "hot" adsorbed species to complete energy equilibration at the substrate temperature (also called thermalization) and momentary trapping within the gas-surface potential (often called sticking if the species remains on the surface for long times, often longer than the measurement). It is often said that the trapped molecule "loses memory" of its initial trajectory after its microscopic motions are scrambled through numerous interactions with surface atoms [19, 20]. These adsorbed molecules may subsequently desorb back into the gas phase (trapping-desorption [21, 22]) at rates that are determined by the surface temperature but by not its initial trajectory or internal states.

We also know that, when the gas-solid or gas-liquid system has come to equilibrium, the outgoing and incoming fluxes of each species must be equal. Langmuir stated this criterion in 1916 with extraordinary prescience: "Since evaporation and condensation are in general thermodynamically reversible phenomena, *the mechanism of evaporation must be the exact reverse of that of condensation*, even down to the smallest detail." [23] In modern terms, a molecular dynamics simulation of gas-solid or gas-liquid collisions can be run backward to simulate the reverse process for every internal [20, 24–26] and velocity component [4, 20] (see water movie at nathanson.chem.wisc.edu by Varilly and Chandler [27]). This microscopic reversibility, a detailed balancing of every molecular process, has an astonishing implication at equilibrium: because the flux of molecules arriving at a surface is Maxwellian and cosine, the flux of molecules leaving the surface must be Maxwellian and cosine too. If the trapping probability depends on incident energy or angle, then the flux of just the desorbing molecules will be non-Maxwellian and non-cosine, with the difference made up by the molecules that directly scatter from the surface! Only the sum of all scattering and desorbing molecules must be Maxwellian and cosine. Thus, if one could observe just the desorbing molecules, one might measure a distribution that is non-Maxwellian and non-cosine, and then infer from it the energy and angular distribution of incoming molecules that undergo trapping and solvation.

A measurement of the desorption distribution can indeed be made in a vacuum experiment (where there is almost no impinging flux) if one assumes that the distribution out of equilibrium in the vacuum chamber is the same as at equilibrium. The history of these concepts for gas-solid interactions is told with great clarity and suspense by Comsa and David [4] and by Kolasinski [20]. I have also learned much from several original references [28–30].

3 Rules of Thumb for Gas-Surface Energy Transfer and Trapping

Three key concepts and examples from gas-surface scattering can be used to appreciate the implications of detailed balancing, as summarized below.

1. The kinematics of the collision govern energy transfer: light gas atoms or molecules bounce off heavy surface atoms or molecules, transferring just a fraction of their translational energy upon collision [31]. Conversely, gas species that are heavier than the surface species (often the case for liquid water) will undergo multiple collisions that lead to efficient energy transfer. For an incoming sphere colliding head-on with an initially stationary sphere (zero impact parameter), the energy transfer is given by $\Delta E/E_{\mathrm{inc}} = 4\mu/(1 + \mu)^2$, where $\mu = m_{\mathrm{gas}}/m_{\mathrm{surf}}$ and $E_{\mathrm{inc}} = 1/2\ m_{\mathrm{gas}}\ c_{\mathrm{inc}}^2$. This equation also models an atom striking a flat cube in a perpendicular direction. When $\mu = 1/4$, 64% of the incident kinetic energy of the gas atom is transferred to the surface atom, while it rises to 89% for $\mu = 1/2$. Numerous experiments verify that energy transfer indeed increases with heavier gas and lighter surface molecules [13, 32–34]. Further studies show that grazing collisions (large impact parameter) transfer less energy, and thermal motions generally decrease the overall energy transfer as well. Sophisticated models of energy transfer have been developed that take into account the shape of molecules [35] and surface and their internal excitation, including the development of a "surface Newton diagram" [18, 36].

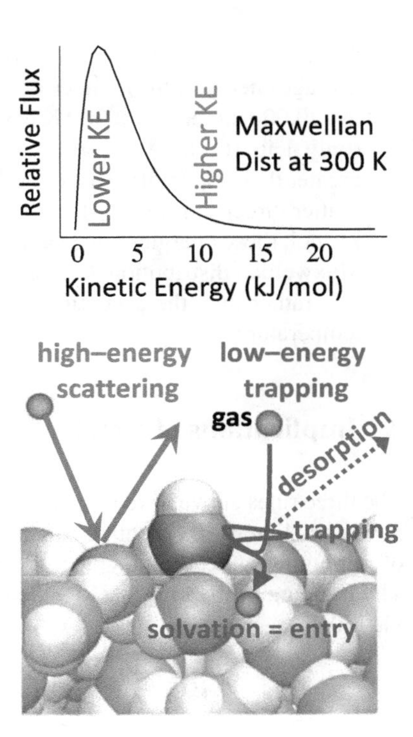

Fig. 2 Two-step mechanism for the dissolution of a gas atom or molecule. In general, high translational energies and grazing collisions lead to direct scattering from the surface, while lower incident energies and more perpendicular collisions lead to energy loss and momentary trapping. This trapping is typically followed by desorption back into the gas phase or diffusion and solvation in the bulk

2. Attractive forces create gas-surface potential energy wells that can momentarily trap the incoming molecule once it has dissipated its excess energy after one or several bounces, as pictured in Fig. 2. For the simple model above, the minimum initial translational energy required to escape the potential energy well is

$$E_{\text{min}} = 4\mu/(1-\mu)^2 \cdot \varepsilon \tag{1}$$

where ε is the well depth [11, 30]. This expression neatly separates into kinematic (mass) and potential energy terms. For $\mu = 1/4$ and $\varepsilon = 20$ kJ/mol (a hydrogen bond between gas and liquid), E_{min} is 36 kJ/mol or 14 RT_{liq} at 300 K—only gases with higher energy will escape thermalization and momentary trapping. Again, experiments verify that heavy gas atoms/light surface atoms and strong attractive forces enhance trapping via the strength of the reagent or product desorption signal [13, 33, 37]. We note the inherent distributions of attractive forces and impact parameters arising from bumpy surfaces, molecular orientation, varying approach angles, and multiple collisions, along with thermal motions of the surface atoms, will broaden the sharp cutoff imposed by Eq. 1. The value of E_{min} might then be taken as midway along the trapping probability curve [20, 30].

3. The Maxwellian flux distribution in terms of translational energy E is given by $J(E) = E/(RT)^2 e^{-E/RT}$. This function peaks at $E = RT = 2.5$ kJ/mol at

300 K and has an average value of 2 RT = 5.0 kJ/mol (not 3/2 RT, which is the average energy of the number-density distribution). In the example above, only 1 in 120,000 molecules at 300 K have translational energies greater than 36 kJ/mol (only 1 in 160 have energies greater than 18 kJ/mol and 1 in 8 have energies greater than 9 kJ/mol). This is a general result: while even heavy gases will often scatter directly from a surface at high collision energies of many 10s to 100s of kJ/mol, these energies have vanishingly low probabilities in a room temperature Maxwellian distribution. Full energy dissipation and trapping (adsorption) is the rule rather than the exception for most molecules on most surfaces near room temperature.

4 Implications of Detailed Balance

The three rules above have immediate implications for the evaporation of gases from solids and liquids. By detailed balancing, the desorption flux J_{des} is equal to the flux of impinging molecules J_{trap} that are momentarily trapped in the interfacial region. The trapping probability $\beta(E, \theta)$ then connects the desorbing and impinging fluxes via [30]

$$\begin{aligned} J_{des}(E,\theta) &= J_{trap}(E,\theta) = \beta(E,\theta)\cdot J_{inc}(E,\theta) \\ &= \beta(E,\theta)\cdot \frac{E}{(RT)^2} e^{-E/RT} \cdot \frac{\cos\theta}{4\pi}\cdot n_{gas} \end{aligned} \quad (2)$$

such that $\beta(E, \theta)$ may be considered both the trapping probability (J_{trap}/J_{inc}) and the evaporation probability (J_{des}/J_{inc}). The rules of thumb above suggest that $\beta(E, \theta)$ will be constant and close to one for most gases, especially when the liquids are made of light molecules such as water (where μ often exceeds one) and where dispersion, dipolar, and hydrogen bonding interactions occur. Thus, most gas molecules should evaporate in a distribution that is close to Maxwellian and cosine at room temperature from water and organic liquids, but perhaps not from solid or liquid metals [30, 38].

Deviations from the typical rules for trapping can reveal underlying mechanisms. One deviation occurs when $\beta(E, \theta)$ changes significantly over the energies in a Maxwellian distribution (0 to ~7 RT_{liq}) or at grazing angles, most likely because of light gas/heavy surface masses (small μ) and weak attractions ε. In these cases, collisions at low energy should lead to trapping while the molecules will scatter at higher energies (as predicted by Eq. 1). Detailed balancing then requires that the adsorbate will desorb in a speed distribution tilted toward lower translational energies because $\beta(E, \theta)$ steadily declines from high to low values as E increases. Rettner and coworkers indeed show this behavior for argon atoms desorbing from hydrogen-covered tungsten, whose sub-Maxwellian desorption matches the distribution of incoming Ar atoms that are momentarily trapped at the surface [30]. This study is mandatory reading for its clarity and precision.

Conversely, imagine an H_2 molecule dissociating upon collision with a metal surface, such as copper. It must have enough translational energy to overcome the ~20 kJ/mol barrier in order to break the H–H bond and form surface Cu–H bonds. High energies along the surface normal facilitate this dissociative adsorption. In the reverse associative desorption, the two adsorbed H atoms come together to generate an H_2 molecule that suddenly finds itself repulsively close to the surface and leaves the surface at high energies, preferentially along the surface normal, that match the incoming energies that lead to dissociation. This detailed balancing of dissociative adsorption and recombinative desorption is observed in pioneering experiments by Cardillo and coworkers and by others [4, 28].

5 Maxwellian Evaporation and a Two-Step Model for Solvation

Now we come to the question in this chapter. Are there also deviations in Maxwellian evaporation from liquids and solutes dissolved in them? During 30 years of observation, we have monitored the vacuum evaporation of liquids such as glycerol, ethylene glycol, alkanes and aromatics, fluorinated ethers, and water from sulfuric acid and pure and salty water itself [39–43]. We have also recorded the evaporation of solute atoms and molecules such as Ar, N_2, O_2, HCl, HBr, HI, Cl_2, Br_2, BrCl, N_2O_5, HNO_3, CO_2, SO_2, $HC(O)OCH_3$, CH_3OCH_3, CH_3NHCH_3, and butanol from one or more of the solvents listed above and others [39, 40, 42, 44–48]. We observed Maxwellian speed distributions in every case (except when the vapor pressure is so high that the gas expands supersonically [41, 49]). This observation is in accord with the arguments above, so we were not surprised. For solvent evaporation and condensation, the mass of the evaporating solvent is necessarily equal to the mass of the surface molecules ($\mu = 1$). In this case, there is very efficient energy transfer (just like billiard balls) and the attractive forces that cohere the molecules into a liquid also trap the gaseous solvent molecule upon collision with the surface. For hydrogen-bonding gases, the attractive forces are also very strong and lead to significant trapping.

We also find, however, that even Ar and N_2 evaporating from salty water evaporate in Maxwellian distributions (within our signal to noise) [42, 43, 50]. By detailed balancing, this Maxwellian evaporation implies that collisions of Ar and N_2 at energies populated in a Maxwellian distribution must thermally equilibrate upon collision. This nearly complete thermalization is likely promoted by the soft nature of surfaces composed of water and organic molecules (it is not true for liquid metals) and weak attractions of a few RT_{liq}.

Separate studies provide insights into the mechanism of dissolution and reaction. Reactive scattering experiments probe HCl → DCl exchange in collisions of HCl with liquid D_2O/D_2SO_4 and of $Cl_2 \rightarrow Br_2$ exchange and $N_2O_5 \rightarrow Br_2$ oxidation in NaBr/glycerol [39, 45, 46]. In all three cases, the product DCl or Br_2 evaporates in a Maxwellian distribution at the temperature of the liquid. The measurements reveal that the ratio of the desorbing product to trapping-desorption (TD) component of the reactant (J(product)/J_{TD}(reactant)) is independent of reactant collision energy from near-thermal to hyperthermal energies. These observations suggest a two-step process for dissolution and reaction (as illustrated in Fig. 2): [39] (1) incoming molecules either directly scatter from the surface or dissipate their excess translational energy, becoming momentarily trapped within the gas-surface potential energy well and losing memory of their initial trajectory, and then (2) these trapped molecules either evaporate or dissolve into the bulk at rates that are determined by gas and liquid properties and temperature, but not by their initial trajectory. In this two-step process, reaction occurs after thermalization within the interfacial region or deeper in the bulk. For the reversible solvation of a non-reacting gas (such as Ar or CH_3OCH_3), evaporation occurs along the reverse pathways, starting with the solute molecule diffusing from the bulk to the surface and then being jettisoned in a Maxwellian distribution into the vacuum by numerous energy-exchanging encounters with surface molecules. This two-step mechanism likely applies to the dissolution and evaporation of most gaseous solutes, but are there exceptions?

6 Non-Maxwellian Evaporation Discovered!

Faubel and Kisters first observed the non-Maxwellian evaporation of acetic acid dimers from a water microjet in 1988, which they attributed to repulsive ejection of the hydrophobic dimer at the surface of water [51]. This single observation persisted until we recorded the non-Maxwellian evaporation of helium atoms in 2014 [50]. Our measurements came about by accident: we were generating microjets [42, 50, 52] of alkane solutions to mimic evaporation of jet fuel in vacuum, which were created by pressurizing a sealed reservoir of the liquid with Ar or N_2 (as first developed by Manfred Faubel [53, 54] and described in Chap. 26). As shown in Fig. 3, the pressurized liquid then emerges from a glass tube with a tapered hole as narrow as 10 μm in diameter. We found that, by vigorously shaking the reservoir, gas can be dissolved into the liquid, which then evaporates as the thin liquid stream exits the nozzle and passes through the vacuum chamber. We have exploited the Maxwellian evaporation of dissolved Ar atoms as "argon jet thermometry" because the Ar speed distribution yields the instantaneous temperature of the jet.

Helium may also be used as a pressurizing gas to create microjets. To our astonishment, we found that He evaporation is non-Maxwellian for every solvent tested, including octane, dodecane, squalane, jet fuel, ethylene glycol, and pure and salty water (as shown in Fig. 4 for 7 M LiBr and 7 M LiCl in water) [42, 43, 50]. Importantly, its behavior is opposite to expectations: instead of evaporating in a slower, sub-Maxwellian distribution, as predicted by argon desorbing from tungsten mentioned above, the He atoms evaporate in a distinctly faster, *super-Maxwellian* distribution! The extent of non-Maxwellian behavior can be gauged by the average translational energy of the exiting He atoms: $1.14 \cdot (2RT_{liq})$ for dodecane at 295 K, $1.37 \cdot (2RT_{liq})$

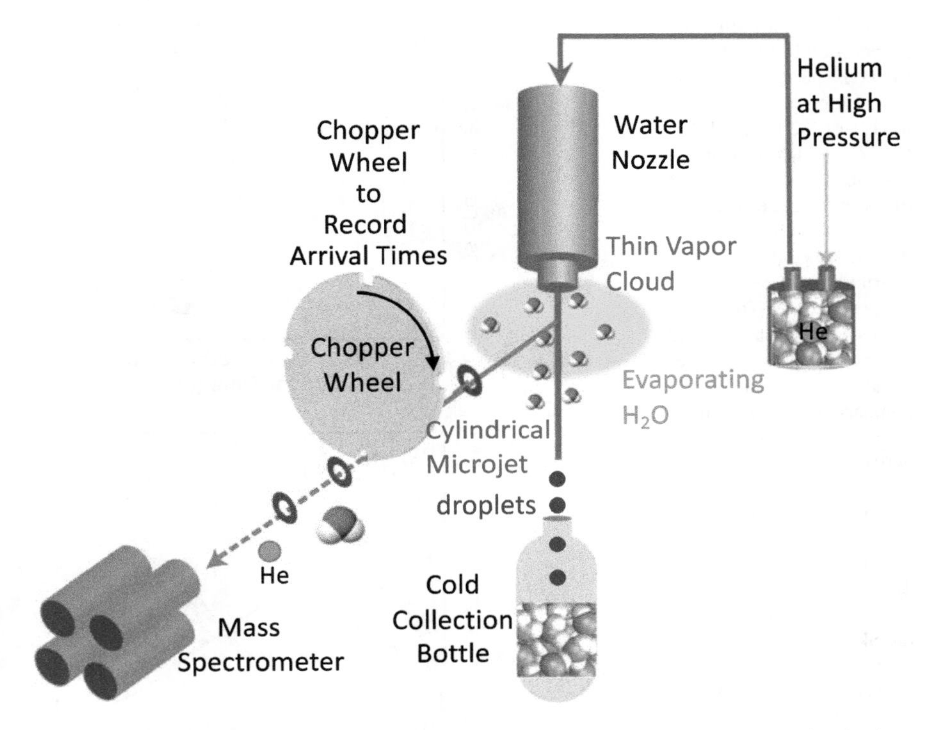

Fig. 3 Vacuum evaporation of helium from pure and salty water microjets. The microjet is a fast-moving thin stream of solvent typically thinner than a strand of hair. When the jet radius is significantly smaller than the He-water mean free path, nearly all He atoms avoid collisions with evaporating water molecules in the vapor cloud surrounding the jet. The jet diameters range from 10 to 35 μm and travel at ~20 m/s. The breakup lengths vary from less than 1 mm for pure water at 252 K to 7 mm for 7 M LiBr/H_2O at 235 K

for pure supercooled water at 252 K, and $1.70 \cdot (2RT_{\text{liq}})$ for 7 M LiBr/H_2O at 255 K, which are 14, 37, and 70% higher than expected [42, 43].

Detailed balancing provides a fascinating interpretation: the super-Maxwellian evaporation of He atoms implies that the reverse process of He dissolution must also be super-Maxwellian.[1] The translational energies of He atoms that dissolve are shifted to higher values, such that the solvation probability, the analog of the trapping probability, increases with increasing collision energy. This result may be interpreted to mean that some He atoms dissolve by "ballistic penetration", pushing water molecules slightly aside as they pass through the interfacial region and enter the liquid! The measured, relative evaporation probabilities $\beta(E)$ for 7 M LiCl and LiBr in water are shown in Fig. 4, which by detailed balance are also the relative solvation

[1]We note that the reverse scattering experiments of He from liquids is complex, involving at least four pathways: direct recoil, trapping-desorption, trapping-dissolution-evaporation, and "ballistic" entry and evaporation. Our attempts to separate the processes have not been successful, as the TOF spectra are dominated by direct recoil. Dissolution appears to be a rare event.

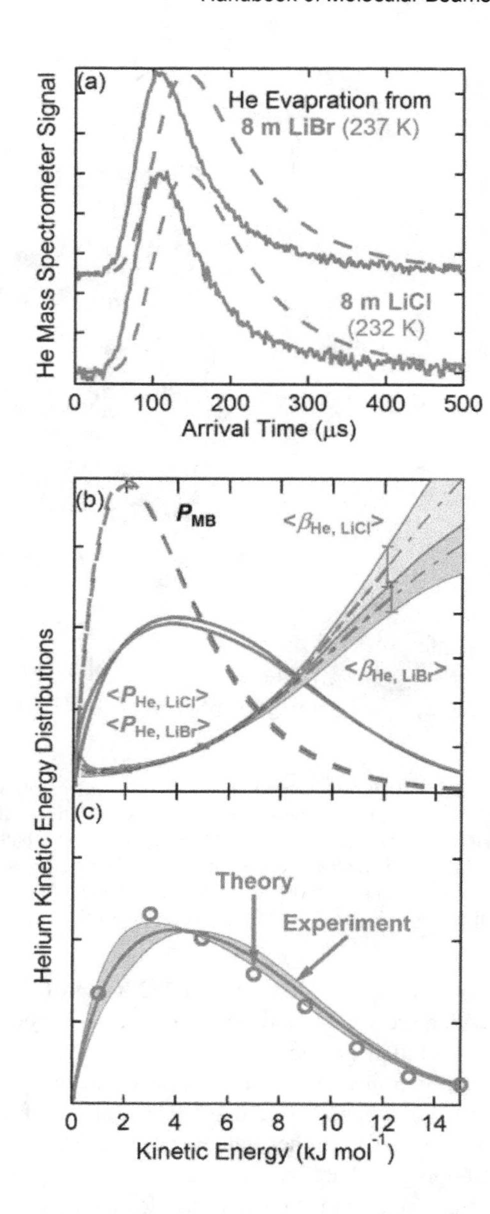

Fig. 4 Examples of helium evaporation from salty water. (**a**) TOF spectra of He atoms evaporating from 8 molal (7 M) LiBr (232 K) and LiCl (237 K), which peak at significantly shorter arrival times (higher speeds and kinetic energies) than the dashed Maxwellian distributions at each temperature. (**b**) The corresponding translational energy distributions of the He atoms, again in comparison to Maxwellian distributions (dashed lines, here called P_{MB}). The relative solvation probabilities $\beta(E)$ (dot-dash) each rise steadily with kinetic energy (see Footnote 2). Panel c shows the excellent agreement between the Skinner/Kann simulations and measurements. This figure is reproduced from Ref. [43]

probabilities for He atoms.[2] Both curves rise steadily with increasing evaporation energy (which is also the collision energy for the reversed trajectories). Many of these He atoms therefore circumvent the two-step trapping-dissolution mechanism

[2]Because we measure only relative He fluxes in the experiments, not absolute fluxes, $J_{des}(E)$ and $J_{inc}(E)$ are each area-normalized in Fig. 4, only ratios of $\beta(E)$ at different E are meaningful. The angular average over $\beta(E, \theta)$ for the cylindrical microjet is described in Refs. [42, 43].

described above as they pass through the interfacial region. But why? Helium atoms have the lowest polarizability of any atom in the periodic Table (0.2 Å^3). In turn, the He-surface potential energy may be so shallow (less than RT_{liq}) that the attractive forces cannot capture He atoms at the surface for the time needed for He atoms to dissolve—thermal motions of the surface molecules instead immediately kick most of the He atoms back into the gas phase. In this case, a substantial fraction of the He atoms cannot enter the liquid via the adsorbed state because the shallow well cannot trap them. We know that the high-energy evaporation of He does not originate from its low mass because the opposite behavior is observed for the evaporation of H_2 from water. For this even lighter gas, evaporation is indeed sub-Maxwellian, as predicted kinematically: energy transfer between H_2 and water is inefficient ($\mu = 0.11$), and only the low energy H_2 molecules lose enough energy to be trapped in the H_2-water potential energy well (H_2 is 4 times more polarizable than He). We also note that the only other gas we have observed that displays super-Maxwellian behavior is neon, which is also weakly polarizable (0.4 Å^3).

7 A View from the Interior

Detailed balancing arguments are beautiful and rigorous and in accord with experiments, but they leave us yearning to know more. How do the dissolved He atoms "know" to evaporate in the same super-Maxwellian distribution that leads to dissolution? It must be so because a single (but complex) potential energy function for all He-water and water-water interactions governs the reverse evaporation and condensation processes [29]. Comsa and David [4] quote an early pioneer, Peter Clausing, who described the detailed balancing requirement of the cosine angular distribution for condensation and evaporation as an "incomprehensible wonder machine", but this statement could apply to the speed distribution as well. My theory colleague Jim Skinner and his student Zak Kann set out to make helium evaporation from water comprehensible, but their explanation is still full of wonder.

Skinner and Kann first performed classical molecular dynamics simulations of He atoms dissolved in pure liquid water [43]. Their simulations indeed show that dissolved He atoms possess a Maxwellian speed distribution right up to the top one to two layers of water, where the He atom is then accelerated into vacuum during the final few collisions of He with H_2O molecules moving outward. My students and I had hoped that the measured He speed distributions would reveal new features of gas-water interactions, but the agreement between simulation and measurement was excellent! Skinner and Kann then went a step further, calculating the Potential of Mean Force (PMF) on the He atom. This potential is equal to the free energy of the He atom as it is dragged infinitely slowly through the interface and into the bulk, sampling all configurations of the water molecules along the way. The free energy curve (PMF) of He in pure water calculated at 255 K is shown in Fig. 5. It starts high in bulk water and decreases to the gas phase value: the difference between the asymptotes is equal to the (very positive) free energy of solvation of approximately

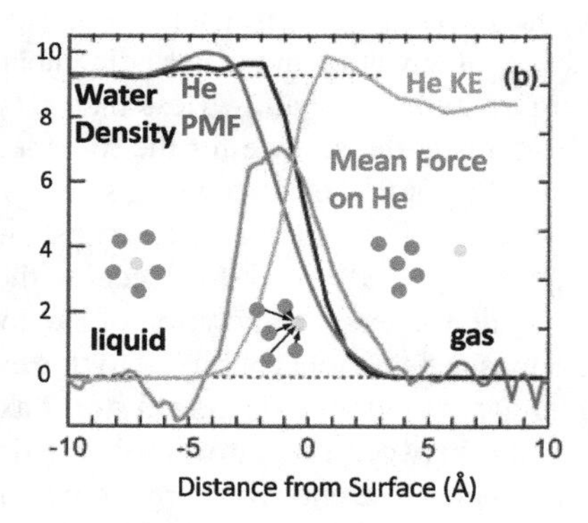

Fig. 5 Potential of Mean Force (PMF) description of an He atom being expelled from water at 255 K. The black curve is the liquid water density, for which the 0 distance is the Gibbs dividing surface. The blue curve is the calculated helium PMF (free energy of solvation), and the green curve is the mean force (negative derivative of the PMF), which spikes in the interfacial region. The grey curve is the resulting helium atom kinetic energy. For these curves, the PMF spans 0 to 10 kJ/mol, the He density-averaged KE spans (3/2)RT = 3.2 kJ/mol to ~1.5×(3/2)RT = 4.7 kJ/mol, and the mean force spans −0.5 to +2.3 kJ/mol/Å. The small drop in He kinetic energy after 1 Å reflects the weak attractive force between He and surface water molecules decelerating the He atom as it leaves. This figure is adapted from Ref. [43]

9.5 kJ/mol at the supercooled 252 K temperature of the 10 μm diameter microjet. Accordingly, helium has the lowest solubility of any gas in water, equal to n_{water}/n_{gas} ~ 1/100 at 252 K. The free energy curve may possess a small barrier (<0.5 kJ/mol) between the surface and bulk regions, but displays at most a very shallow minimum at the surface. This is unlike even N_2 or O_2, which are also weakly soluble but whose attraction to H_2O generate weak adsorption wells (>2 kJ/mol) [55].

Why then do He atoms emerge at higher than Maxwellian translational energies? The negative derivative of the free energy curve is just the "mean force" associated with the PMF—it is the repulsive force acting on the He atom itself as it moves infinitely slowly through the liquid! Figure 5 shows this mean force spikes right at the interface where the He atoms are accelerated [43]. Here is the key point: if the He atom indeed moved slowly through the interfacial region, it would undergo enough energy-exchanging collisions with water molecules at each point to maintain a Maxwellian distribution. But the He atoms do not move slowly, and at some point they stop equilibrating as the interfacial density becomes sparser (as in Fig. 5) and there are insufficient He–H_2O collisions to absorb the extra He atom energy. In this case, the He atom "detaches" from the PMF and exits into vacuum, carrying its excess energy with it imparted by the repulsive forces. In a sense, the water molecules "squeeze" the interloping He atom into vacuum as the water-water hydrogen bonds "heal" to their native structure.

We note that the PMF only describes the force perpendicular to the surface. Skinner and Kann have also investigated the angular distributions of evaporating atoms, and deduce that the perpendicular component is even more super-Maxwellian but is partially canceled by sub-Maxwellian parallel components [56]. This study includes a wide-ranging investigation of the effects of solute mass and solute-solvent attractive forces on solute evaporation, including confirmation that H_2 is sub-Maxwellian and Ne is super-Maxwellian. Parallel simulations by Williams, Patel, and Koehler of He evaporation from dodecane lead to a fascinating "cone and crater" mechanism by which He atoms are expelled in an exposed cone at the surface whose walls may crater inward, accelerating the He atom from the cone [57, 58].

One rule of thumb emerges from these investigations: the more insoluble the gas, the steeper the PMF, the greater the force on the evaporating gas atoms, and the more likely that the He atom will emerge in a non-Maxwellian distribution. Thus, higher He atom exit energies should accompany lower solubilities in different solvents, a trend that we observe experimentally [43]. This correlation is not quantitative, however, because the PMF describes a slowly moving solute atom that fully equilibrates as it moves through solution and samples all configurations of the water molecules—it is the breakdown of this picture arising from insufficient He-water collisions in the outermost region that gives rise to an excess kinetic energy. A focus on the mean force and interfacial collisions instead provides an exquisite statistical framework that can guide future investigations.

8 Future Non-Maxwellian Adventures

What are some potential new directions for helium evaporation experiments? The demonstration of super-Maxwellian He evaporation is the closest we have come to He atom diffraction from periodic solid surfaces. The question of what can be learned from He scattering from liquids was one my students and I asked when we began in 1988, and it took until now to address it: super-Maxwellian He evaporation from liquids reflects the forces acting on the He atom in the outermost layers of the liquid. Skinner's and Kann's successful simulations [43] suggest that He evaporation from pure and salty water may not contribute to a refined picture of gas-water interactions because they were already so successful in replicating the energy distributions. But water is almost never pure or even just salty. Oceans, lakes, aerosol particles, and tap water contain numerous organic species, many of which are surface active [59–61]. We hope in future studies to investigate surfactant-coated microjets prepared with soluble ionic species such as tetrabutylammonium bromide and neutral ones such as butanol or pentanoic acid [47]. Helium evaporation from these surfactant solutions may reveal how gases move through loosely to tightly packet alkyl chains, depending on their bulk-phase concentration, and thus provide information on the mechanisms of gas transport through monolayers [62]. It will also be intriguing to mimic the seminal studies of the Cardillo and Comsa groups [4, 28], who investigated H_2 permeation and desorption through metals. We can monitor the parallel evaporation of He atoms through thin polymer films of functionalized organic polymers and even self-assembled monolayers over a wide range of exit angles. It is inspiring to imagine that Stern might have enjoyed these studies, an extension of his "method of molecular rays" to liquids in vacuum, "for which I [Stern] consider the directness and simplicity as the distinguishing property."

Acknowledgements This work was supported by the National Science Foundation (CHE-1152737) and the Air Force Office of Scientific Research. I thank the many contributions of my students working on the projects described here, including Alexis Johnson, Diane Lancaster, Jennifer Faust, Christine Hahn, and Tom Sobyra, who all became masters of microjets and committed untold hours to making these experiments work. I am also indebted to Zak Kann and Jim Skinner for a bountiful collaboration and for teaching us the extraordinary insights that statistical mechanics can provide. I am grateful to Dudley Herschbach for introducing me to the life of Otto Stern and to Dudley and Peter Toennies for being Honorary Chairpersons of the Otto Stern conference. My special thanks go to Bretislav Friedrich and Horst Schmidt-Böcking for organizing and directing every aspect of the conference, which spanned science, history, philosophy, music, science outreach, and a dazzling excursion on the Rhein River. Otto Stern leapt off the page through loving tributes, history lessons, and the astonishingly manifold applications of the molecular ray technique that he founded. We all glimpsed a multidimensional human being—a fantastically insightful and fearless scientist and outstandingly moral mentor in dark times. His words and deeds have become a constant companion.

References

1. O. Stern, A direct measurement of thermal molecular speed. Z. Phys. **2**, 49–56 (1920)
2. O. Stern, Addition to my work "a direct measurement of thermal molecular speed". Z. Phys. **3**, 417–421 (1920)
3. I. Estermann, O.C. Simpson, O. Stern, The free fall of atoms and the measurement of the velocity distribution in a molecular beam of cesium atoms. Phys. Rev. **71**, 238–249 (1947)
4. G. Comsa, R. David, Dynamical parameters of desorbing molecules. Surf. Sci. Rep. **5**, 145–198 (1985)
5. A.H. Persad, C.A. Ward, Expressions for the evaporation and condensation coefficients in the Hertz-Knudsen relation. Chem. Rev. **116**, 7727–7767 (2016)
6. J.C. Maxwell, Illustrations of the dynamical theory of gases—Part I. On the motions and collisions of perfectly elastic spheres. London, Edinburgh Dublin Phil. Mag. J. Sci. **19**, 19–32 (1860)
7. J.C. Maxwell, VII. On stresses in rarified gases arising from inequalities of temperature. Phil. Trans. **170**, 231–256 (1879)
8. I. Estermann, O. Stern, Diffraction of molecular beams. Z. Phys. **61**, 95–125 (1930)
9. J.A. Barker, D.J. Auerbach, Gas-surface interactions and dynamics; thermal energy atomic and molecular beam studies. Surf. Sci. Rep. **4**, 1–99 (1985)
10. S.T. Ceyer, New mechanisms for chemical at surfaces. Science **249**, 133–139 (1990)
11. C.T. Rettner, D.J. Auerbach, J.C. Tully, A.W. Kleyn, Chemical dynamics at the gas-surface interface. J. Phys. Chem. **100**, 13021–13033 (1996)
12. G.M. Nathanson, Molecular beam studies of gas-liquid interfaces. Ann. Rev. Phys. Chem. **55**, 231–255 (2004)
13. J.W. Lu, B.S. Day, L.R. Fiegland, E.D. Davis, W.A. Alexander, D. Troya, J.R. Morris, Interfacial energy exchange and reaction dynamics in collisions of gases on model organic surfaces. Prog. Surf. Sci. **87**, 221–252 (2012)
14. H. Chadwick, R.D. Beck, Quantum state resolved gas-surface reaction dynamics experiments: a tutorial review. Chem. Soc. Rev. **45**, 3576–3594 (2016)
15. M.A. Tesa-Serrate, E.J. Smoll, T.K. Minton, K.G. McKendrick, Atomic and molecular collisions at liquid surfaces, in *Ann. Rev. Phys. Chem.*, vol. 67, ed. by M.A. Johnson, T.J. Martinez (2016), pp. 515–540
16. C.H. Hoffman, D.J. Nesbitt, Quantum state resolved 3D velocity map imaging of surface scattered molecules: incident energy effects in HCl plus self-assembled monolayer collisions. J. Phys. Chem. C **120**, 16687–16698 (2016)

17. F. Zaera, Use of molecular beams for kinetic measurements of chemical reactions on solid surfaces. Surf. Sci. Rep. **72**, 59–104 (2017)
18. W.A. Alexander, Particle beam scattering from the vacuum-liquid interface, in *Physical Chemistry of Gas-Liquid Interfaces*, ed. by J.A. Faust, J.E. House (Elsevier, The Netherlands, 2018), pp. 195–234
19. C.T. Rettner, D.J. Auerbach, Distinguishing the direct and indirect products of a gas-surface reaction. Science **263**, 365–367 (1994)
20. K.W. Kolasinski, *Surface Science: Foundations of Catalysis and Nanoscience*, 3rd edn. (Wiley, United Kingdom, 2012). Ch. 3
21. J.E. Hurst, C.A. Becker, J.P. Cowin, K.C. Janda, L. Wharton, D.J. Auerbach, Observation of direct inelastic scattering in the presence of trapping-desorption scattering—XE ON Pt(111). Phys. Rev. Lett. **43**, 1175–1177 (1979)
22. M.E. Saecker, G.M. Nathanson, Collisions of protic and aprotic gases with hydrogen-bonding and hydrocarbon liquids. J. Chem. Phys. **99**, 7056–7075 (1993)
23. I. Langmuir, The constitution and fundamental properties of solids and liquids part I solids. J. Am. Chem. Soc. **38**, 2221–2295 (1916)
24. D.F. Padowitz, S.J. Sibener, Sublimation of nitric oxide films—rotation and angular distributions of desorbing molecules. Surf. Sci. **217**, 233–246 (1989)
25. H.A. Michelsen, C.T. Rettner, D.J. Auerbach, R.N. Zare, Effect of rotation on the translational and vibrational energy dependence of the dissociative adsorption of D_2 on Cu(111). J. Chem. Phys. **98**, 8294–8307 (1993)
26. M.J. Weida, J.M. Sperhac, D.J. Nesbitt, Sublimation dynamics of CO_2 thin films: a high resolution diode laser study of quantum state resolved sticking coefficients. J. Chem. Phys. **105**, 749–766 (1996)
27. P. Varilly, D. Chandler, Water evaporation: a transition path sampling study. J. Phys. Chem. B **117**, 1419–1428 (2013)
28. M.J. Cardillo, M. Balooch, R.E. Stickney, Detailed balancing and quasi-equilibrium in adsorption of hydrogen on copper. Surf. Sci. **50**, 263–278 (1975)
29. J.C. Tully, Dynamics of gas-surface interactions: thermal desorption of Ar and Xe from platinum. Surf. Sci. **111**, 461–478 (1981)
30. C.T. Rettner, E.K. Schweizer, C.B. Mullins, Desorption and trapping of argon at a 2H–W(100) surface and a test of the applicability of detailed balance to a nonequilibrium system. J. Chem. Phys. **90**, 3800–3813 (1989)
31. E.K. Grimmelmann, J.C. Tully, M.J. Cardillo, Hard-cube model analysis of gas-surface energy accommodation. J. Chem. Phys. **72**, 1039–1043 (1980)
32. M.E. King, G.M. Nathanson, M.A. Hanninglee, T.K. Minton, Probing the microscopic corrugation of liquid surfaces with gas-liquid collisions. Phys. Rev. Lett. **70**, 1026–1029 (1993)
33. M.E. Saecker, G.M. Nathanson, Collisions of protic and aprotic gases with a perfluorinated liquid. J. Chem. Phys. **100**, 3999–4005 (1994)
34. B.G. Perkins, D.J. Nesbitt, Quantum-state-resolved CO_2 scattering dynamics at the gas-liquid interface: Incident collision energy and liquid dependence. J. Phys. Chem. B **110**, 17126–17137 (2006)
35. T.Y. Yan, W.L. Hase, J.C. Tully, A washboard with moment of inertia model of gas-surface scattering. J. Chem. Phys. **120**, 1031–1043 (2004)
36. W.A. Alexander, J.M. Zhang, V.J. Murray, G.M. Nathanson, T.K. Minton, Kinematics and dynamics of atomic-beam scattering on liquid and self-assembled monolayer surfaces. Faraday Discuss. **157**, 355–374 (2012)
37. T.B. Sobyra, M.P. Melvin, G.M. Nathanson, Liquid microjet measurements of the entry of organic acids and bases into salty water. J. Phys. Chem. C **121**, 20911–20924 (2017)
38. M. Manning, J.A. Morgan, D.J. Castro, G.M. Nathanson, Examination of liquid metal surfaces through angular and energy measurements of inert gas collisions with liquid Ga, In, and Bi. J. Chem. Phys. **119**, 12593–12604 (2003)
39. J.R. Morris, P. Behr, M.D. Antman, B.R. Ringeisen, J. Splan, G.M. Nathanson, Molecular beam scattering from supercooled sulfuric acid: collisions of HCl, HBr, and HNO_3 with 70 wt % D_2SO_4. J. Phys. Chem. A **104**, 6738–6751 (2000)

40. A.H. Muenter, J.L. DeZwaan, G.M. Nathanson, Collisions of DCl with pure and salty glycerol: enhancement of interfacial D → H exchange by dissolved NaI. J. Phys. Chem. B **110**, 4881–4891 (2006)
41. S.M. Brastad, G.M. Nathanson, Molecular beam studies of HCl dissolution and dissociation in cold salty water. Phys. Chem. Chem. Phys. **13**, 8284–8295 (2011)
42. D.K. Lancaster, A.M. Johnson, K. Kappes, G.M. Nathanson, Probing gas–liquid interfacial dynamics by helium evaporation from hydrocarbon liquids and jet fuels. J. Phys. Chem. A 14613–14623 (2015)
43. C. Hahn, Z.R. Kann, J.A. Faust, J.L. Skinner, G.M. Nathanson, Super-Maxwellian Helium evaporation from pure and salty water. J. Chem. Phys. 144 (2016)
44. B.R. Ringeisen, A.H. Muenter, G.M. Nathanson, Collisions of HCl, DCl, and HBr with liquid glycerol: gas uptake, D → H exchange, and solution thermodynamics. J. Phys. Chem. B **106**, 4988–4998 (2002)
45. L.P. Dempsey, J.A. Faust, G.M. Nathanson, Near-interfacial halogen atom exchange in collisions of Cl_2 with 2.7 M NaBr-glycerol. J. Phys. Chem. B. **116**, 12306–12318 (2012)
46. M.A. Shaloski, J.R. Gord, S. Staudt, S.L. Quinn, T.H. Bertram, G.M. Nathanson, Reactions of N_2O_5 with salty and surfactant-coated glycerol: interfacial conversion of Br- to Br 2 mediated by alkylammonium cations. J. Phys. Chem. A **121**, 3708–3719 (2017)
47. T.B. Sobyra, H. Pliszka, T.H. Bertram, G.M. Nathanson, Production of Br_2 from N_2O_5 and Br^- in salty and surfactant-coated water microjets. J. Phys. Chem. A **123**, 8942–8953 (2019)
48. T. Krebs, G.M. Nathanson, Reactive collisions of sulfur dioxide with molten carbonates. Proc. Natl. Acad. Sci. U.S.A. **107**, 6622–6627 (2010)
49. D.K. Lancaster, A.M. Johnson, D.K. Burden, J.P. Wiens, G.M. Nathanson, Inert gas scattering from liquid hydrocarbon microjets. J. Chem. Phys. Lett. **4**, 3045–3049 (2013)
50. A.M. Johnson, D.K. Lancaster, J.A. Faust, C. Hahn, A. Reznickova, G.M. Nathanson, Ballistic evaporation and solvation of helium atoms at the surfaces of protic and hydrocarbon liquids. J. Phys. Chem. Lett. **5**, 3914–3918 (2014)
51. M. Faubel, T. Kisters, Non-equilibrium molecular evaporation of carboxylic acid dimers. Nature **339**, 527–529 (1989)
52. J.A. Faust, G.M. Nathanson, Microjets and coated wheels: versatile tools for exploring collisions and reactions at gas-liquid interfaces. Chem. Soc. Rev. **45**, 3609–3620 (2016)
53. M. Faubel, S. Schlemmer, J.P. Toennies, A molecular beam study of the evaporation of water from a liquid jet. Z. Phys. D. **10**, 269–277 (1988)
54. M. Faubel, Photoelectron spectroscopy at liquid surfaces, in *Photoionization and Photodetachment: Part I*, vol. 10A, ed. by C.-Y. Ng (World Scientific, Singapore, 2000), pp. 634–690
55. R. Vácha, P. Slavíĉek, M. Mucha, B.J. Finlayson-Pitts, P. Jungwirth, Adsorption of atmospherically relevant gases at the air/water interface: free energy profiles of aqueous solvation of N_2, O_2, O_3, OH, H_2O, HO_2, and H_2O_2. J. Phys. Chem. A **108**, 11573–11579 (2004)
56. Z.R. Kann, J.L. Skinner, Sub- and super-Maxwellian evaporation of simple gases from liquid water. J. Chem, Phys **144**, 154701 (2016)
57. S.V.P. Koehler, M.A. Williams, MD simulations of he evaporting from dodecane. Chem. Phys. Lett. **629**, 53–57 (2015)
58. E.H. Patel, M.A. Williams, S.P.K. Koehler, Kinetic energy and angular distributions of He and Ar atoms evaporating from liquid dodecane. J. Phys. Chem. B **121**, 233–239 (2017)
59. R.E. Cochran, O. Laskina, J.V. Trueblood, A.D. Estillore, H.S. Morris, T. Jayarathne, C.M. Sultana, C. Lee, P. Lin, J. Laskin, A. Laskin, J.A. Dowling, Z. Qin, C.D. Cappa, T.H. Bertram, A.V. Tivanski, E.A. Stone, K.A. Prather, V.H. Grassian, Molecular diversity of sea spray aerosol particles: impact of ocean biology on particle composition and hygroscopicity. Chem **2**, 655–667 (2017)
60. T.H. Bertram, R.E. Cochran, V.H. Grassian, E.A. Stone, Sea spray aerosol chemical composition: elemental and molecular mimics for laboratory studies of heterogeneous and multiphase reactions. Chem. Soc. Rev. **47**, 2374–2400 (2018)
61. K. Jardak, P. Drogui, R. Daghrir, Surfactants in aquatic and terrestrial environment: occurrence, behavior, and treatment processes. Environ. Sci. Pollut. Res. **23**, 3195–3216 (2016)
62. G.T. Barnes, Permeation through Monolayers. Colloids Surf. A **126**, 149–158 (1997)

4

Ultra-Fast Dynamics in Quantum Systems Revealed by Particle Motion as Clock

M. S. Schöffler, L. Ph. H. Schmidt, S. Eckart, R. Dörner, A. Czasch, O. Jagutzki, T. Jahnke, J. Ullrich, R. Moshammer, R. Schuch and H. Schmidt-Böcking

Abstract To explore ultra-fast dynamics in quantum systems one needs detection schemes which allow time measurements in the attosecond regime. During the recent decades, the pump & probe two-pulse laser technique has provided milestone results on ultra-fast dynamics with femto- and attosecond time resolution. Today this technique is applied in many laboratories around the globe, since complete pump & probe systems are commercially available. It is, however, less known or even forgotten that ultra-fast dynamics has been investigated several decades earlier even with zeptosecond resolution in ion-atom collision processes. A few of such historic experiments, are presented here, where the particle motion (due to its very fast velocity) was used as chronometer to determine ultra-short time delays in quantum reaction processes. Finally, an outlook is given when in near future relativistic heavy ion beams are available which allow a novel kind of "pump & probe" experiments on molecular systems with a few zeptosecond resolution. However, such experiments are only feasible if the complete many-particle fragmentation process can be imaged with high momentum resolution by state-of-the-art multi-particle coincidence technique.

M. S. Schöffler · L. Ph. H. Schmidt · S. Eckart · R. Dörner · A. Czasch · O. Jagutzki · T. Jahnke · H. Schmidt-Böcking (✉)
Institut für Kernphysik, Universität Frankfurt, 60348 Frankfurt, Germany
e-mail: hsb@atom.uni-frankfurt.de; schmidtb@atom.uni-frankfurt.de

A. Czasch · O. Jagutzki · H. Schmidt-Böcking
Roentdek GmbH, 65779 Kelkheim, Germany

J. Ullrich
PTB, Brunswick, Germany

R. Moshammer
MPI für Kernphysik, Heidelberg, Germany

R. Schuch
Physics Department, Stockholm University, 107 67 Alba Nova, Stockholm, Sweden

1 Introduction

To explore the nature of atomic matter scientists have developed during the last century sophisticated approaches to reveal the microscopic structure of matter and also the dynamics between atoms or even inside atoms and molecules. The resolving power for static structural features of molecular systems, e.g. measured by Cryo-electron microscopy [1] or X-ray spectroscopy [2], is presently in a range of a few 10^{-10} m, which is about a few times the diameter of a single atom. In these measurement approaches the momenta of electrons or photons scattered on a molecular object are detected. The measured momentum distributions, are converted by Fourier transformation into coordinate space, yielding a spatial image of the molecular structure.

To explore the dynamics of a reaction between quantum objects or to reveal the electron dynamics inside a quantum object the experimenter in general interacts with a fast projectile (photon, electron, ion etc.) in a first step (the excitation or ionization step) with the quantum object and observes after very short time delays (typically attoseconds) electron and ionic fragment emission. Thus, the experimenter obtains information on the dynamically changed final states or even intermediate states of the object.

In order to reveal the "entangled" electron dynamics inside the same molecule, it is typically not sufficient to perform single parameter measurements on the same molecular object at two shortly successive instants in time. To reveal entangled dynamics, the simultaneous detection of the momenta of all fragments emitted from the same single molecule is required. Thus, a multi-fragment coincidence measurement imaging the complete momentum space with high resolution is necessary. Such high-resolution multi-coincidence detection systems are available since about two decades: The COLTRIMS-reaction microscope C-REMI possesses all necessary properties to perform such high-resolution multi-coincidence investigations [3].

What are the time delays Δt of interest? The duration of a chemical reaction is typically in the order of a picosecond, and, accordingly, a nucleus can be considered as locally frozen during a time interval of a femtosecond [4]. Therefore, to explore such nuclear processes a time resolution of about one femtosecond (10^{-15} s) is required. This is the standard time range where modern femtosecond Laser pump-and probe-schemes can visualize chemical reactions, geometrical changes and their dynamics [5]. Intra-atomic or intra-molecular processes do proceed faster. The duration of a charge transfer process in ion-atom or ion/molecule collisions depends on the projectile velocity and on the quasi-molecular promotion path-way [6]. It can be as short as an attosecond (10^{-18} s). Hole migration in photon-excited molecules can proceed also in the attosecond range, as well as intra-atomic or intra-molecular Coulombic vacancy or energy transfer. Interatomic (or intermolecular) electronic decay processes occur in a wide range of durations from few femtoseconds to several picoseconds [7]. Energy transfer processes as in photosynthesis of chlorophyll, for example, proceed on the upper femtosecond level.

In fast ion-atom collisions intra-atomic and intra-molecular dynamics can take place even on the lower zeptosecond level (10^{-19} s), which is about 4 orders of magnitude shorter than the typical femtosecond Laser pulse can resolve [8–12]. In such a short time interval light travels only a distance of 0.3 Å. To visualize these ultra-fast dynamics, one needs detection methods which visualize its time dependence, e.g. in interference structures like in quantum-beats. As will be shown below when dynamical processes proceed via two different pathways they accumulate different phases yielding characteristic interference structures. From these structures phase differences can be determined and, as outlined before, by knowing the velocity of the fast ion, time delays even in the zeptosecond regime can be deduced [9].

It may be a more theoretical and philosophical issue whether ultra-short time scales below one attosecond may be of any relevance in atomic physics. But these time scales are doubtlessly of high interest in quantum physics, in general. For example, fundamental questions arise, as whether the so-called "collapse" of a wave function is a local process and starts in one location inside a molecule and proceeds then with speed of light through the whole molecular system. In this case the "collapse" would last about 300 zeptoseconds to stretch across a simple molecule. Or is the collapse a non-local process instantaneously present everywhere across the molecule? Measurements with 10 zeptosecond time resolution would allow to explore such a fundamental question e.g. in a triatomic molecule with a non-linear geometry (see Sect. 3.3).

2 Ultra-fast Chronometer Mechanisms Using Fast Moving Particles as Clock

Burgdörfer et al. [8] have recently presented a review on the historic development and the present status of attosecond physics performed in the field of ion-atom collisions and short-pulse Laser physics. By discussing the theoretical aspects, they have shown the similarity of ion- and Laser-induced processes. Since the chronometer scheme of the multi-photon pump & probe technique is discussed widely in [5, 8, 12], this paper will concentrate on ionization processes induced by ion impact, where the motion of particles provides the ultra-fast chronometers.

The method of "pumping" a quantum object to an excited or ionized state and "probing" this excited state, i.e. by observing the delayed fragment or photon emission, is in general the principle of any measurement in reactions between quantum particles*. So-called "Pump & Probe" measurements are today commonly identified with Two-pulse Laser Pump & Probe methods where the very short time delay $\mathbf{\Delta t}$ between the two Laser pulses can be well adjusted by two different geometrical path ways yielding a time resolution in the femto- or even attosecond range. In this Laser Pump & Probe approach the probing is processed via a delayed second photon pulse where the delay time can be chosen by the experimenter.

In ion-atom collisions the experimenter can never prepare two projectiles ions such that they interact with an atom or molecule at the same impact parameter with a well-controlled time delay of attosecond precision to undergo like in Laser physics a Pump & Probe process. The "pump & probe" process in ion-atom collisions must be induced by the same ion at two different locations in a molecule. Since the relative locations of atoms in a molecule are known with a precision of about 0.1 a.u., the delay-time between the ion reaction at two different locations can then be varied by changing the ion velocity—and typically achievable ion velocities correspond to 10 zeptosecond-pump-probe-resolution.

2.1 Historic Life-Time Measurements with Nano- and Picosecond Precision

Measurements with time-delay determination have been performed already 100 years ago, e.g. by Stern and Volmer [13], when they measured the mean decay time of photon-excited I_2 molecules (Fig. 1). Since many of the articles in these proceedings of the Otto-Stern conference are related to Otto Stern's scientific work we will shortly discuss here Stern's and Volmer's pioneering work of measuring life times of excited molecules, too. Stern and Volmer used the thermal motion of vapor molecules as chronometer. The excited molecules expanded from a tiny interaction spot of a few micrometer diameter, where they have been excited, according to their thermal motion (i.e. the motion of the molecules created streaking). Stern and Volmer observed the excitation and decay positions of the molecules using a light microscope. From the outreaching tails of the excitation spot Stern and Volmer derived the

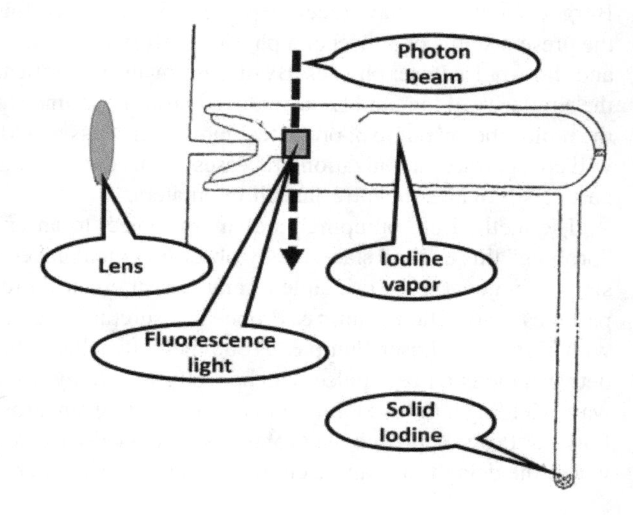

Fig. 1 Inside the glass tube I_2 molecules were evaporated from solid Iodine. A very narrow collimated photon beam (1 μm diameter) excited the molecules (small quadratic box). The fluorescence light emission was observed in a greater halo region due to the thermal motion of the excited molecules. This halo distribution was measured using a lens system. From the halo distribution and the thermal properties, the mean decay lifetime was determined [13]

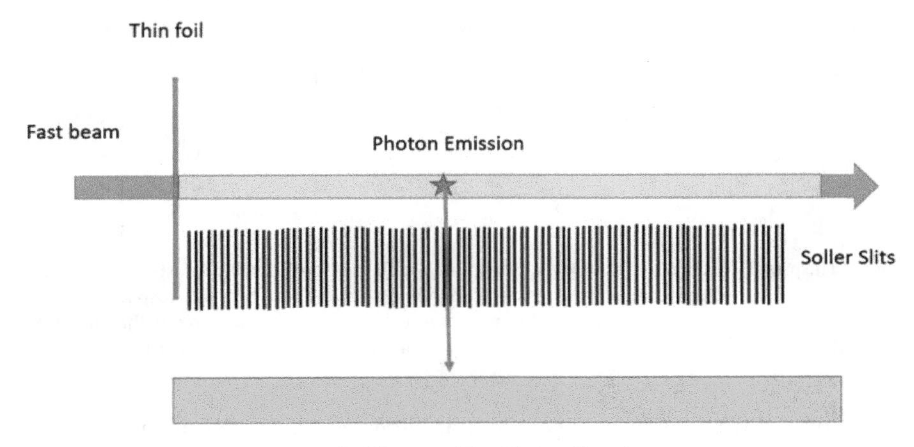

Fig. 2 Beam foil spectroscopy with fast ions [14]

mean life time for the decay process of the excited states with a time resolution of about 2 ns.

In the past, numerous methods have been developed and applied to measure decay times and explore dynamics in atomic and molecular systems. For historic reasons we describe here also the so-called "beam-foil" techniques [14]. A fast ion beam (kinetic energy typically 0.1 to 10 MeV/u) penetrates a very thin foil. Inside the foil ions get excited and decay downstream the moving beam. The emitted fluorescence photons are detected with a position-sensitive photon detector. A Soller-slit system allows only photons emitted transversely to reach the detector.

Thus, the exponential decay distribution as function of distance from the foil (i.e. decay time) is measured. From the exponential slope one can calculate the delay time with picosecond precision ("beam-foil" techniques see Fig. 2) [14]. This is a kind of streaking technique where from the observable positions (foil and decay) and from the ion velocity the decay time is deduced by macroscopic methods.

If inside the foil two nearby ionic levels can be excited simultaneously, the time-resolved fluorescence light emitted from these coherently excited levels can show a characteristic quantum beat structure.

2.2 *Quantum Beat Structures as Ultra-fast Chronometers*

Ultra-short time interval measurements can be performed also with fast moving particles. As "clock" the fast motion of ions is used whose trajectory can be considered as classical. Measuring transition probabilities resulting from two spatially localized interaction areas the two transitions amplitudes interfere yielding a characteristic interference pattern. Since the delay-time can be calculated from the classical motion

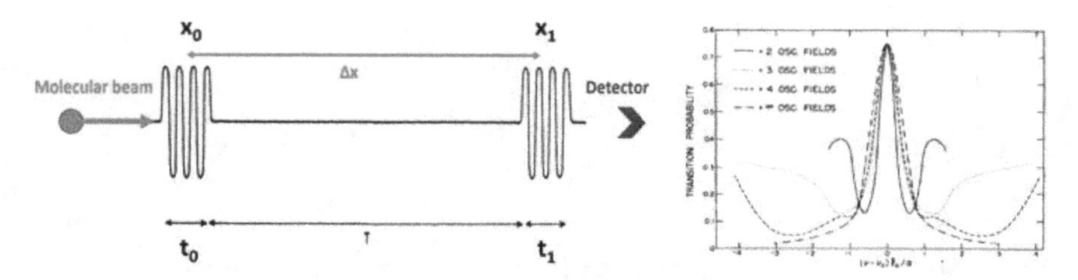

Fig. 3 The principle scheme of Ramsey's "Separated oscillating field" device. A fast-moving object (here indicated as molecular beam) passes two coherent cavities at time t_0 and t_1 respectively, and the object can be excited at time t_0 and t_1. From the interference structure measured in the excitation probability (see right side) the phase difference $\Delta\Phi$ of both amplitudes can be determined [15]

of the ion and from the locations where transitions occur, the phases can be determined from the measured interference pattern and thus information on the dynamics of the reaction process can be derived.

This superposition scheme of two wave amplitudes for moving atoms emitted at time t_0 and t_1 (see Fig. 3) was already applied in Ramsey's "Separated oscillating field method", which is the basis of the atomic clock [15]. In Ramsey's pioneering experiment a moving object (in Fig. 3 indicated as molecular beam) passes through two cavities and the moving object can be excited either at t_0 or t_1. Since the experimenter does not know in which cavity the excitation took place both excitation amplitudes at t_0 and t_1 add coherently. From the interference structure measured in the excitation probability (see right side of Fig. 3) the phase difference $\Delta\Phi$ of both amplitudes can be determined. From $\Delta\Phi$ and the known time delay T the transition energy can be deduced with high precision.

3 Experimental Examples of Quantum-Beat Measurements in Ion-Atom/Molecule Collisions

3.1 Quantum Beats in Quasi-molecular X-Ray Emission

In specially prepared ion-atom collision processes one can use the fast classical motion of an ion as a very fast clock to visualize even electronic dynamics with a time resolution in the lower zeptosecond regime. By measuring the quantum beat structure in the spectra of quasi-molecular X-ray emission Schuch et al. [9] have obtained in fast ion-atom-collisions even a time resolution of nearly 10 zeptoseconds. The X-rays emitted in a reaction visualize the streaking of the quasi-molecular orbitals by the two-center ion-atom nuclear potential, which provides a very fast, with time varying streaking force. The X-ray photon energies encode, thus, the fingerprints of the strength of the streaking force at the moment of emission and yield information on Δt.

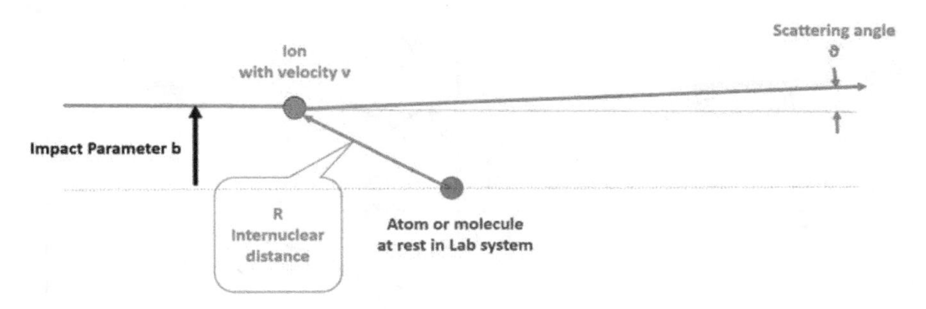

Fig. 4 Scheme of an ion-atom scattering process

The purpose of the experiment by Schuch et al. [9] was to measure the binding energy of the very short living quasi-molecular 1sσ-state which was formed in fast ion-atom collisions for a time duration of only about 1 attosecond. More than 30 years ago several groups (quasi-molecular radiation) [10] and (K-K vacancy sharing) [11] have measured quantum beat structures in ion-atom collisions with oscillations in the atto- and zeptosecond range. From the interference structures of these quantum beats, phase differences were determined yielding energy or time domain information. In Fig. 4 the scheme of such an ion-atom scattering process is presented. An X-ray/scattered-projectile coincidence measurement is required to reveal such quantum-beat structures for a given impact parameter. To probe such a short living quasi-molecular state with the detection techniques of the eighties was extremely difficult since the achieved quasi-molecular X-ray/scattered-projectile coincidence rate was a few true counts per hour. In the laboratory system the projectile ion bypasses an atom on a quasi-straight line (very small deflection angles of about 1°, which are determined by detecting the scattered projectile deflection angle). From the deflection angle the impact parameter b can be deduced. Since the ion velocity is known the internuclear distance R (vector) can be calculated as function of the relative collision time t ($t_0 = 0$ is the time moment at distance of closest approach).

If R is much smaller than the projectile ion or target atom K-shell radii even the most-inner electronic orbitals steadily approach during this extremely short (sub-attosecond) collision time the united-atom electronic states due to the combined projectile and target nuclear Coulomb potentials. Thus, the combined nuclear potentials "streak" as function of $R = R(t)$ the energy values of the bound quasi-molecular states (see Fig. 5).

To reveal the streaked quasi-molecular energy values, one has to prepare the projectile in a very special ionic configuration to create observable quantum beat structures. If one bombards a hydrogen-like Cl^{16+} projectile ion on an Ar atom, thus, a 1sσ vacancy is already present on the incoming part of the collision and identical X-rays can be emitted on the way into the collision (−t values) and on the way out of the collision (+t values). Thus the transition amplitudes on the first half (way into the collision) interfere with those of the second half of the collision. Like in a double slit experiment (see Ramsey's atomic-clock [15] "separated-oscillating-field method")

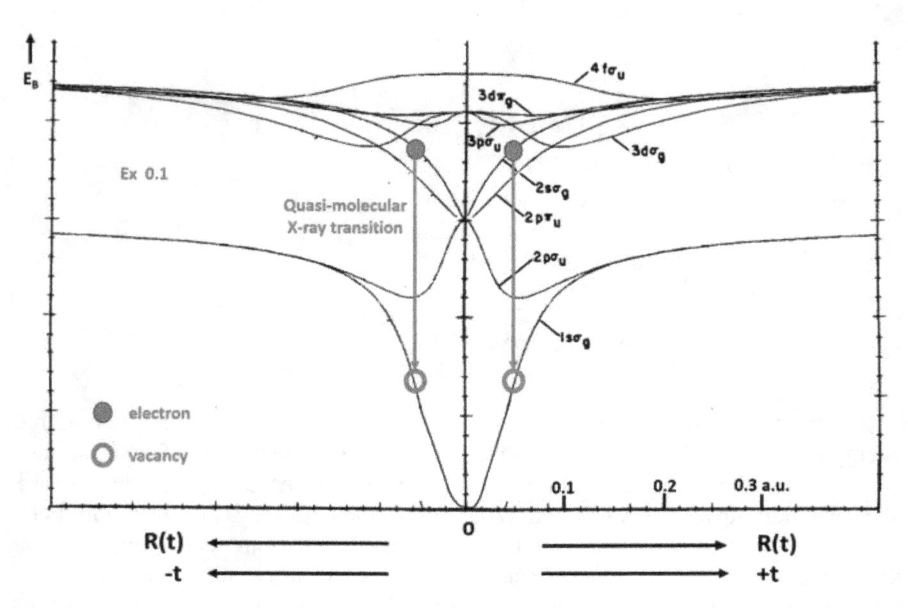

Fig. 5 Quasi-molecular correlation diagram for Cl^{16+} on Ar as function of the internuclear distance R. During the collision an electron from the 2pπ orbital can pass over into the 1sσ orbital and an X-ray is emitted. The X-ray transition energy is the energy difference between the 2pπ and the 1sσ orbital [9] at the particular R value

one does observe characteristic quantum beat structures in the spectra of quasi-molecular X-rays emitted during the collision (see Fig. 6). Since the quantum beat structures vary with impact parameter the X-rays must be detected in coincidence with scattered projectiles to select one given scattering angle (i.e. a fixed impact parameter-range). The X-ray transition process observed here is the quasi-molecular K_α-transition (electron transition from the 2pπ into the 1sσ quasi-molecular state). The K_α-transition energy in the united-atom limit at very small internuclear distances ($Z_{UA} = Z_{Bromium} = Z_{Cl} + Z_{Ar} = 35$) is about 12 keV (see Fig. 7). These X-rays are emitted per definition at the time moment t_0. X-rays of lower energy are emitted at larger internuclear distances R, i.e. larger -t or +t values (see Fig. 5). Thus, the collision time parameter $t = R(t)$ is zero at E_x of the united atom ($E_x = 12$ keV) and increases to larger R values, i.e. lower X-ray energies.

From Fig. 7 we derive that each X-ray transition energy corresponds to a well-defined internuclear distance and thus to a well-defined collision moment. Thus, we can visualize the variation of the quasi molecular Coulomb potential with nearly 10 zeptosecond resolution. Furthermore, the observed quantum beat structure yields a phase information with $\varphi(t) =_{t=0} \int^t E_x(R(t))/\hbar \cdot dt$. The highest X-ray energies correspond to R_{min} or $t = 0$. According to [9] one can now—vice versa—determine the transition energies $E_x(R(t))$ as function of the collision time, i.e. the internuclear distance R. The final result is shown in Fig. 7, where from all measured spectra

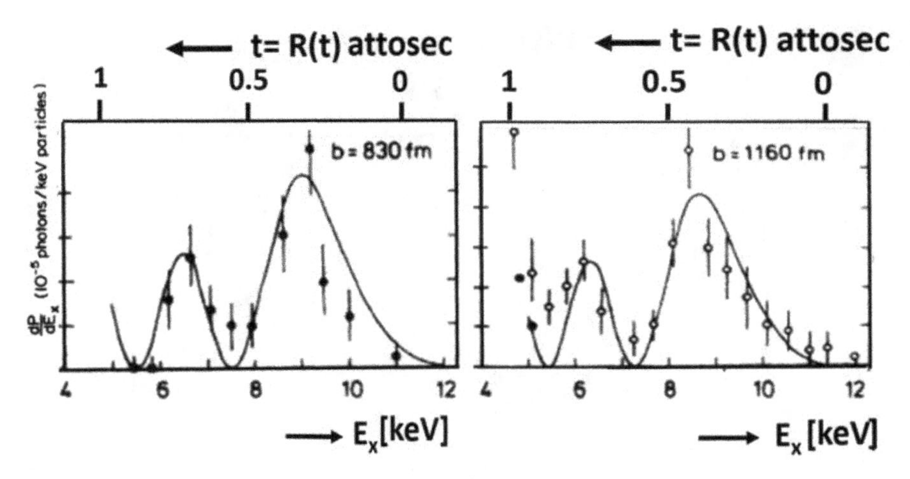

Fig. 6 Measured X-ray spectra for the 2pπ − 1sσ quasi-molecular transition in Cl^{16+}-Ar collisions [9] at fixed impact parameters. The X-ray energy is directly measured with a Si(Li)-detector and can be transferred into an internuclear distance R via the correlation diagram (Fig. 5). From the internuclear distance R and the ion velocity the time scale can be calibrated

(different impact parameters and different collision energies) the quasi-molecular energy values as function of the internuclear distance are displayed.

The analysis of the quantum beats shows that for E_x as function of R an universal curve is obtained independent of the ion-atom collision energy, i.e. independent of the streaking time. For 20 MeV collision energy the R scale (from $R = 0$ up to 0.1 a.u.) in Fig. 7 corresponds to 500 zeptoseconds, for 5 MeV to 1 attosecond. Such measurements [9–11] show that using ion-atom collisions the dynamics of quasi-molecular states could be explored with 10 zeptosecond resolution.

3.2 *Young-Type Interference Structures in Slow $H_2{}^+$ +He Collisions*

Schmidt et al. have [6] investigated "Young-type interference structures" in 10 keV $H_2{}^+ + He \Rightarrow H_2^* + He^+ \Rightarrow H + H + He^+$ collisions (relative velocity $v = 0.45$ a.u.). They measured the momentum vectors of all reaction fragments in the final state in coincidence using the C-REMI approach. Thus, the orientation of the $H_2{}^+$ molecule with respect to the projectile momentum vector was determined for each event yielding the He^+ scattering distribution in dependence of the angle θ (Fig. 8). In the moving projectile system, the He atom can be scattered by both $H_2{}^+$ projectile nuclei (double slit) (see Fig. 8). The scattered He^+ momentum wave function is then the coherent sum of the two amplitudes emerging from the two H nuclei scattering centers. From the resulting interference structures in the He^+ scattering distribution

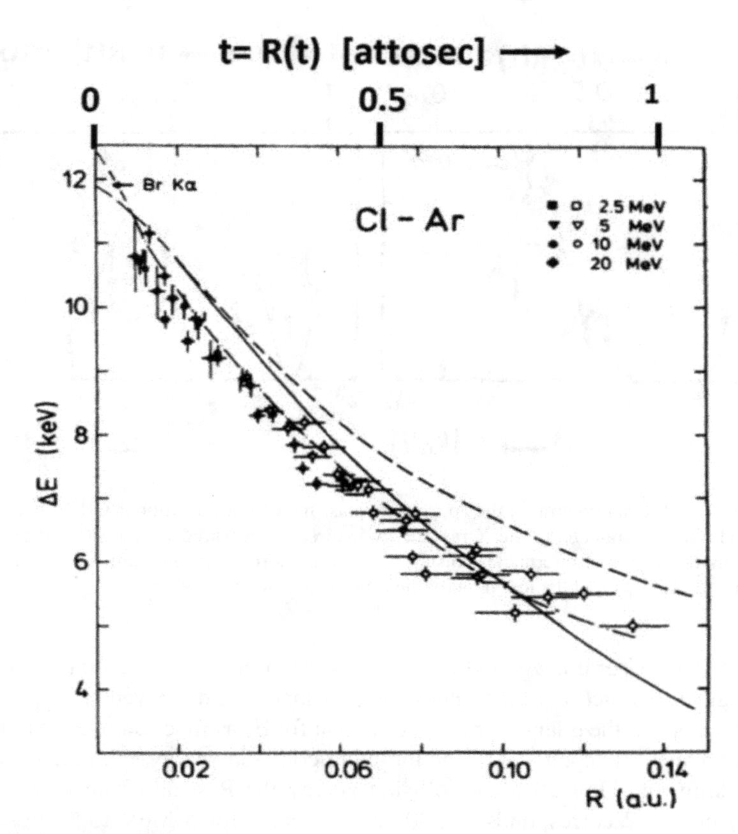

Fig. 7 Transition energies between 1sσ and 2pπ quasi-molecular states as function of the internuclear distance R. The dashed-dotted line results from DFS calculations for the 1sσ − 2pπ transition; solid and dashed lines are scaled from $H^+ + H$ for the 2pσ − 1sσ and 2pπ − 1sσ transition, respectively [9]. The time scale corresponds to 5 MeV collision energy

the phase shifts between the two amplitudes can be deduced as function of Θ visualizing the tiny time delay between the interaction of the He atom with the first (t_1) and second (t_2) H atom (see Fig. 8).

The phase difference between both amplitudes due to the molecular orientation θ is proportional to the measurable delay time ($\Delta t = t_2 - t_1$). In a multi-particle coincidence measurement one can also calculate it directly from the measured angle θ and from the ion velocity v.

Figure 9 clearly shows that the interference pattern varies with θ and even with the KER value (electronic excitation), too. Schmidt et al. presented a model calculation (red dashed line in Fig. 10) for the superposition of the scattering amplitudes at the two H atoms. The differential scattering cross sections can be expressed as $d\sigma/d\varphi \sim \cos^2(\beta/2)$ where the phase shift ß is $ß = \pi + R \cdot \Delta p_{He}/\hbar + \Delta E \cdot \Delta t/\hbar$. The phase jump π accounts for the inversion of the molecular symmetry in the electronic transition

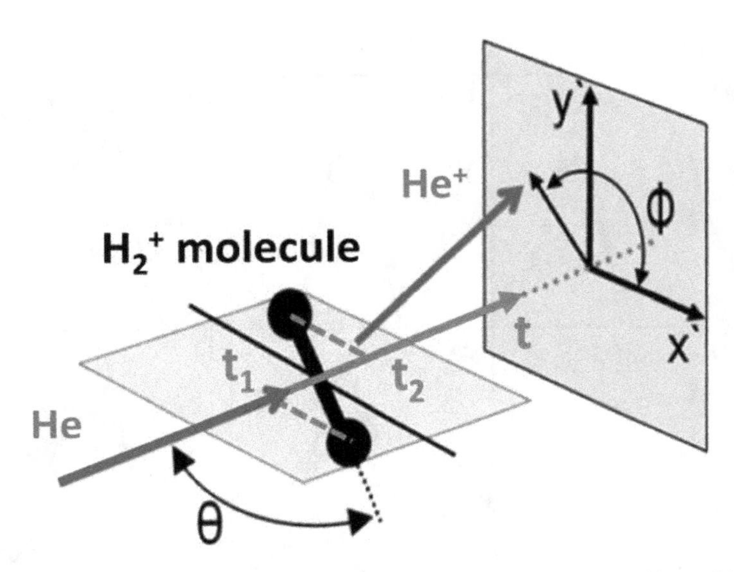

Fig. 8 Nuclear scattering scheme. In inverse kinematics the He target atom collides with the H_2^+ molecular ion and transfers one electron to the H_2 molecule which breaks up into neutral H atoms. The two H atoms are detected at small scattering angles in forward direction. The He^+ ion is detected with a C-REMI under 90°. The He is scattered into the azimuthal angle Φ. The relative orientation of the molecule to the He impact direction is defined by the angle Θ. The time difference $t_2 - t_1$ can be determined from the measured angle Θ and the ion velocity

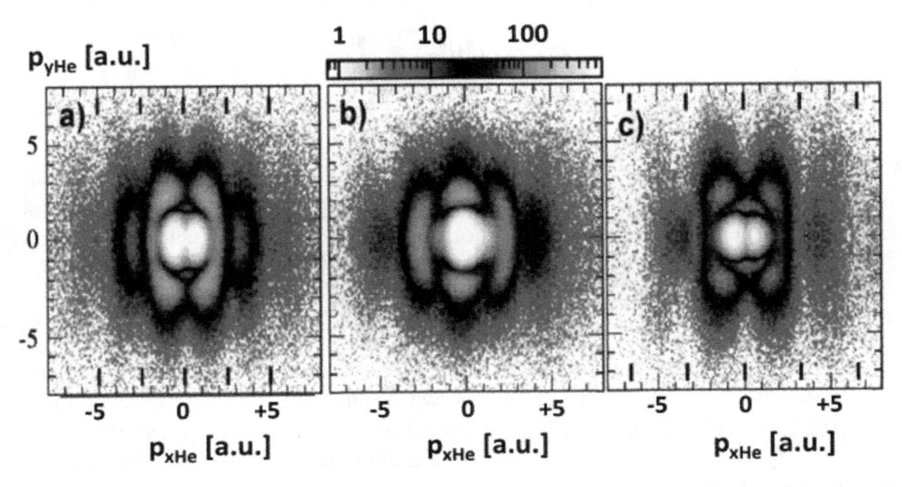

Fig. 9 Two-slit interference pattern in the plane perpendicular to the projectile momentum vector for three different θ angles. **a** Events for molecular orientation angles (with respect to the beam direction) from 80° to 90° and KER* values 1 to 2 eV. This KER corresponds to R values from 2.3 to 2.9 a.u. **b** Events for molecular orientation angles from 50° to 60° and KER 3 to 4 eV. This KER corresponds to R values from 2.3 to 2.9 a.u. **c** Events for molecular orientation angles from 80° to 90° and KER 3 to 4 eV. This KER corresponds to R values from 1.7 to 2.0 a.u. [6]

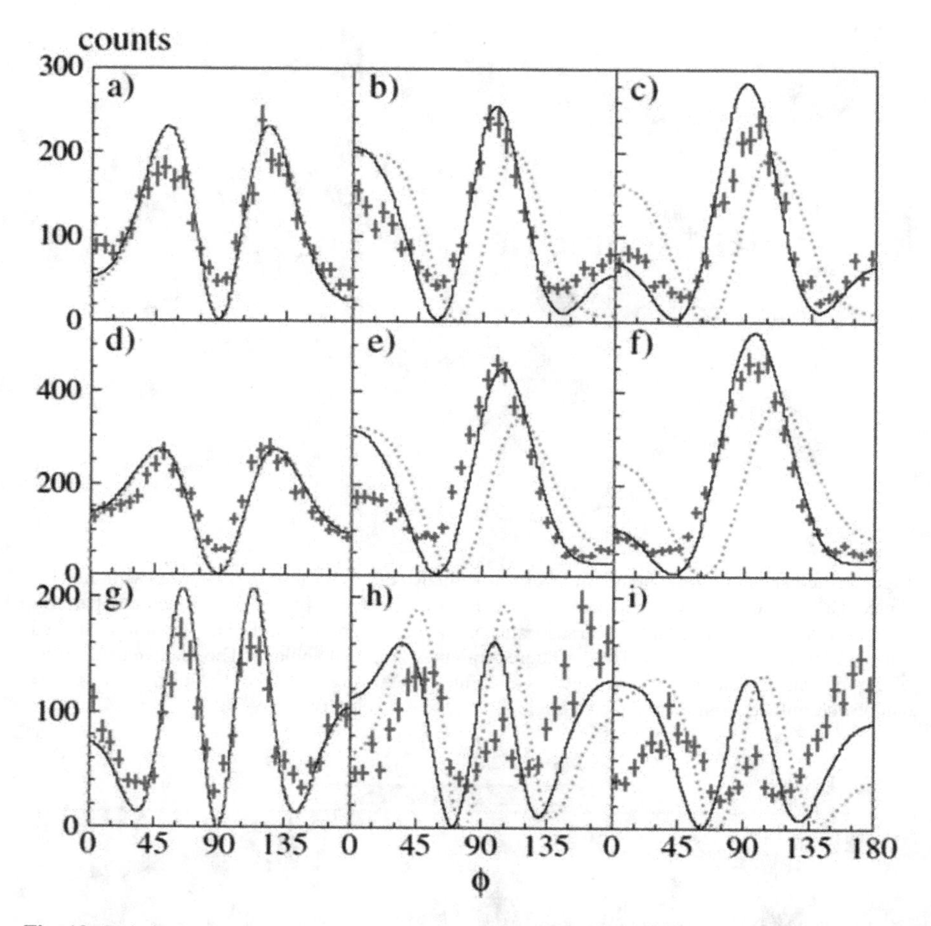

Fig. 10 Relative cross sections as function of the angle φ in comparison with model calculations (red dashed lines) and full theory (solid lines) [6]. The left column represents collisions where θ is 85–90°, the middle column where θ is 55–60° and the right column where θ is 45–50°. The rows show data for different transverse momenta and slightly different KER values

and the second term for the change of the He momentum Δp_{He} due to the scattering. which leads to a change of the de Broglie wave length of the scattered He and for the term $\Delta E \cdot \Delta t/\hbar$. It accounts for the correction of the so-called translation factor with $\Delta t = t_2 - t_1$ (see Fig. 8) as time difference of the interaction of the He projectile with the two H atoms

In Fig. 10, the measured interference pattern are compared with these model calculations (red dashed lines) and the full theory (black lines) [6]. The data are presented for a given polar scattering angle (i.e. impact parameter) as function of the azimuthal angle Φ. In the left column (a + d + g) of Fig. 10 the data are shown for $\theta = 90°$ and $\Delta t = 0$. In the middle column (b + c + h) $\theta = 60°$ and the right column (c + f + i) data are shown for $\theta = 45°$ where the influence of the delay Δt on the scattering as well as electronic excitation becomes strongly visible. The data of Schmidt et al. [6] prove in a convincing manner that the electronic excitation processes as function of the delay Δt does vary.

The full theory (only for the phase variation, not including electronic dynamics for transition probabilities) describes rather well (besides g and h) the measured phase variations. The model calculations (red dashed lines) do not include the nuclear Δt phase effect. They are shifted by about 20° to higher angles and indicate the large effect of Δt on the phases. The good agreement in phase of the full theory proves that the time delay calculated from the geometry and ion velocity are properly taken into account. The time delay Δt (derived from the collision geometry, see Fig. 8) is for θ angles between 45° to 50° $\Delta t = R \cdot \sin(45°)/v_{He} = 2.3$ a.u. $\cdot \sin(45°)/0.45$ a.u. $= 3.6$ a.u. $= 87$ attosecond. The disagreement in the absolute height shows, that the electron dynamics varies with Δt too. The measured data contain also information on the electon dynamics in such reactions.

The limits of the resolution for Δt determined from the experimentally observed phase shifts can be estimated from the data of Fig. 10 and from the comparison with the theoretical calculations. The resolution in determing phases is about 3°. This corresponds for the 10 keV He on H_2 collision system to 10 attosec time resolution in such collisions. If the ion velocity increases the ion-motion based clock would gain resolution. The He on H_2 collision system investigated by Schmidt et al. [6] clearly demonstrates that the effect of time delay between both scattering amplitudes is nicely visible in the interference structure. The absolute scattering intensities in the different final excitation states of the two H atoms are, however, only in modest agreement with the data. It is to notice, that only such channels were measured by Schmidt et al. where the final H fragments remained in the ground state. To observe more significant differences in the excitation of both atoms of the dipolar molecule one should investigate molecular species with higher Z values.

Since the overall momentum resolution is so excellent, the different channels of electronic excitation in these scattering processes can be resolved event by event, as well, and one can identify different electronic promotion channels during the collision. From Fig. 11 one can deduce that for each event the different electron promotion channels (different molecular orbitals (see Fig. 12 and 13)) with KER and Q value are fully separated. The different electronic promotion pathways are marked by the letters "a" to "f" and "A" plus "B". In the three-body (He and two proton nuclei) scattering process the He^+ ion is mostly scattered out of plane into the angle Φ.

In the 10 keV $H_2{}^+ + He \Rightarrow H_2^* + He^+ \Rightarrow H + H + He^+$ electron transfer process one He electron is captured to metastable H_2^* vibrational states (H(1s) + H(2l)) state with an energy minimum at about R = 2 a.u. [6]. Since the internuclear distance R is a function of time, the H_2^* fragmentation process provides a fast clock, too. Thus, one can estimate that the time period for the capture (in the He-H_2 system) lasts only about a few hundred attoseconds. In this reaction channel a fraction of these states $He^+ = > H + H + He^+$ decays after the collision during the ongoing Coulomb explosion (duration about some tenth of femtoseconds) into the H(1s) + H(1s) ground state by

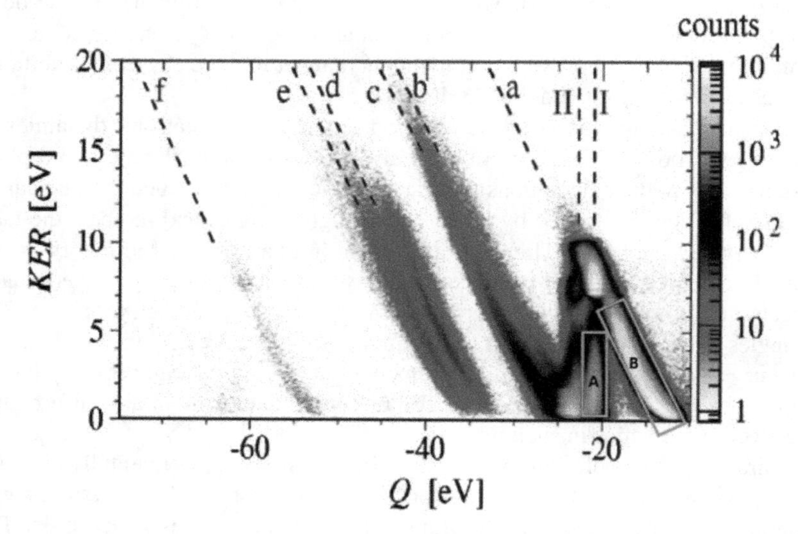

Fig. 11 Final state KER-Q-value correlation diagram of the different electronic excitation channels. The lines a to f mark different exited states. The channels A and B are discussed in more detail in the next Figs. 12 and 13 [6]

Fig. 12 Electron promotion and fragmentation scheme of region B in Fig. 11 [6]

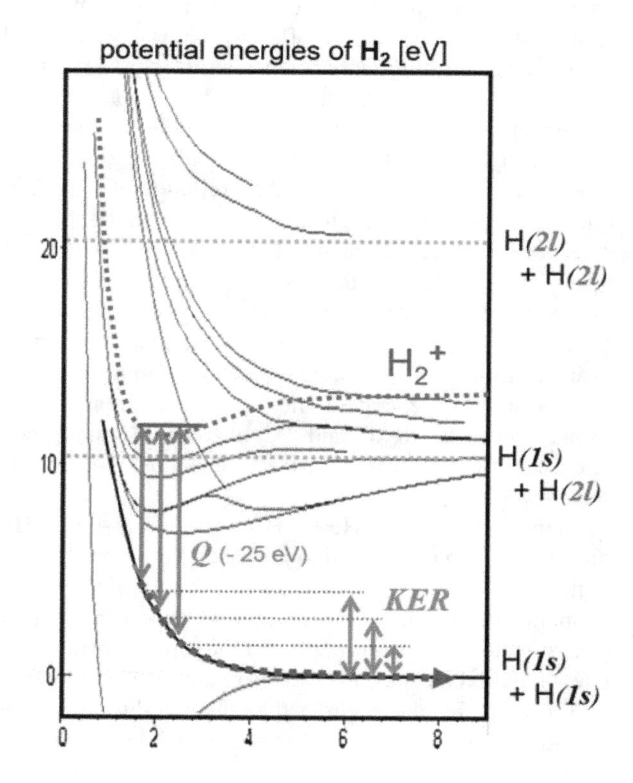

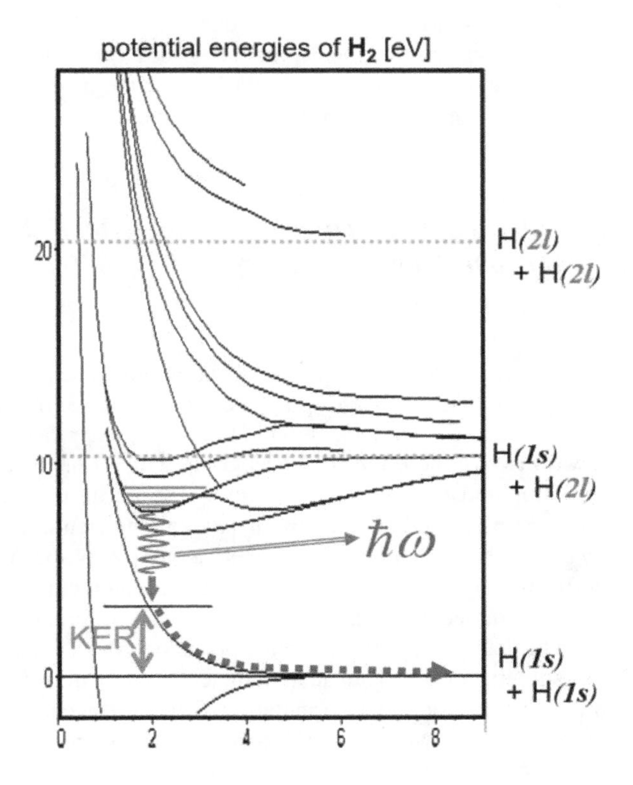

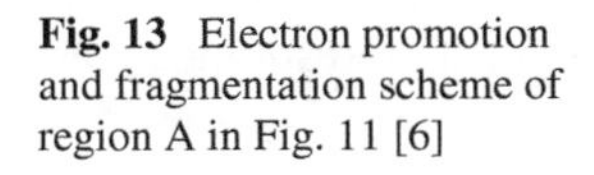
Fig. 13 Electron promotion and fragmentation scheme of region A in Fig. 11 [6]

emitting photons (green oscillation ħω in Fig. 13) with a variable KER energy (see channel A in Fig. 11). The remaining small amount of energy [relative to the ground state H(1s) + H(1s)] is detectable as final KER value.

In the collision 10 keV H_2^+ on He, one He electron can also be captured to the H(1s) + H(1s) ground state (red arrows in Fig. 12) (region B in Fig. 11) [6]. Like for region A, this electron transfer occurs during the ongoing Coulomb explosion into the H(1s) + H(1s) ground state. The remaining amount of energy (relative to the ground state H(1s) + H(1s)) converts to KER in the final state.

Similar experiments for molecules with higher Z-values have been performed recently by Iskandar et al. [16]. Iskandar et al. investigated collisions of low energy Ar^{9+} ions on Ar_2 dimer targets and measured all ionic fragments in coincidence. From the measured recoil-ion momenta the dimer orientation with respect to the projectile direction, the nuclear transverse momentum transfer (impact parameter) and the KER values could be measured for each event. They found clear evidence that the capture from the first hit atom in the dimer is favored. Because of the large distance of both atoms in the dimer the highly charged ion interacts preferable only with one atom with a high probability of multiple capture. The subsequent intra-molecular vacancy sharing probability between the two Ar atoms in the dimer is low because of the large distance of both atoms in the dimer. They analyzed their data in the "Over-barrier Model" and found reasonable agreement between theory and experiment describing the capture processes. Since their momentum resolution was not good enough to distinguish quasi-molecular promotion path-ways they were not able to explore the fast electronic dynamics.

3.3 A Proposal: Scheme of an Ion-Atom/Molecule Pump & Probe Technique Approaching 10 Zeptoseconds Time Resolution

As already pointed out above the ion-atom/molecule pump & probe scheme has little in common with the Laser pump & probe scheme, in which two Laser pulses can be created as an interlocked pair. No experimenter can produce an ion beam where always two of these ions move as group with a time delay adjusted to 1 attosecond precision and passing through a molecule on an identical trajectory. Even if one could prepare such an ion pair, these two ions would immediately repel each other by their nuclear Coulomb force. Thus, a pump & probe process with ions can only be performed by a single (solely) moving ion, interacting at two different locations in the same molecule. These locations (e.g. two different atoms in the same molecule) must be detectable with subatomic precision. Thus, the experimenter must be able to measure the position (impact parameters => transferred momenta) of the projectile-molecule reaction and the orientation of the ion trajectory with respect to the structure of the molecule. Measuring all ionic fragment momenta in the final state by a multi-coincidence-approach with high momentum resolution allows for a deduction of both—the orientation and the impact parameters. Thus, the experimenter knows in which time sequence and time delays the ion interacted with the different atoms in the molecule. E.g. using the C-REMI approach these requirements can indeed be satisfied.

Since the relative distances between atoms in a molecule can be calculated with about 0.1 a.u. precision, and as a fast ion (particularly a relativistic moving heavy ion) follows a perfect classical straight-line trajectory, the relative time delays between the impact at the two different atoms in a molecule can be determined with a few zeptoseconds resolution. But how can one utilize this kind of timing to investigate electronic dynamics in ion-molecule collisions?

An ion moving with a relativistic velocity (v_{ion} => c speed of light) interacts with atoms or molecules via a very sharp retarded dipole-like electric field where the opening angle scales with $1/\gamma$, where $\gamma = 1/\mathrm{sqrt}(1 - (v/c)^2)$. For $\gamma = 20$ the opening angle is about 20° yielding a short interaction time in the lower zeptosecond regime. As shown in [17] the virtual photon field of relativistic heavy ions as projectiles is very strong. It interacts simultaneously with nearly all electrons in the molecule resulting in multiple ionization of the ionic fragments. To control the electron dynamics in such a collision, the momenta of all ejected electrons must be measured in coincidence, too. Thus, the scheme of an ion induced pump & probe measurement presented here

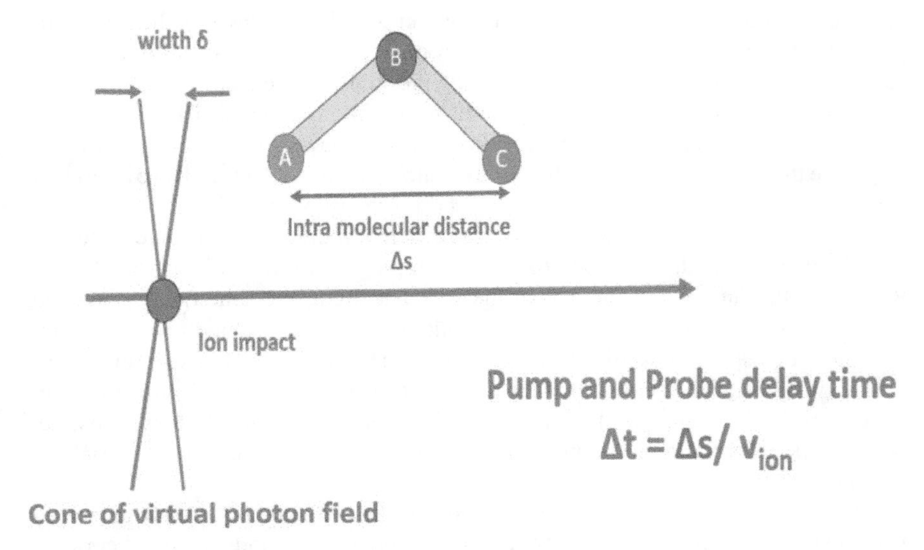

Fig. 14 Ion-molecule pump & probe scheme

requires a multi-coincidence detection approach which must have a high detection efficiency and excellent momentum resolution for ions and electrons.

In Fig. 14 that scheme of an ion-molecule pump & probe measurement is shown. The direction of the impacting ion beam is precisely prepared, however, the molecules (occurring in the gas phase) are randomly oriented.

As an example a triatomic molecule is considered here, where the different atoms A, B, and C are bound in a non-linear formation. The distances between the atoms in the molecule are typically in the order of 2 to 3 a.u. In the example shown here, an ion travelling with the speed of light, interacts first with atom A (pump process) and induces the "collapse" of the molecular ground state wave function. After a time delay of about $\Delta t = \Delta s/c \approx 3$ a.u./137 a.u. ≈ 0.022 a.u. ≈ 480 zeptoseconds the same ion approaches atom C. One interesting question is, when does atom C "know" that atom A suffered the collapse of its wave function? If the collapse is instantly present everywhere in the molecule then atom C is very likely in an excited state, when the ion interacts after about 460 zeptosecond with the molecule at position C. If, however, the collapse emerges from atom A with the speed of light, then the transport of collapse information via atom B to atom C will arrive at best at the same time, but most likely a little later than the relativistic projectile ion. Thus, the observed final state of fragment C may depend on the collapse expansion time.

To explore the nature of the collapse expansion one has to measure the momenta of all ionic fragments and all emitted electrons in the final state in coincidence for all orientations of the molecule with respect to the projectile ion flight direction. A C-REMI provides for each fragment detection a nearly 4π solid angle with about 50% detection efficiency for each fragment. Thus, the total multi-coincidence detection efficiency is rather high (for three ionic fragments and 3 electrons is the total coincidence efficiency still about 2%). However, this multi-particle detection efficiency can be increased dramatically, by optimizing the single particle detection efficiency: enhancing the transmission and open area of the micro channel plates up to 90% [18]. The overall final state of the reaction process may strongly depend in which sequence the ion interacted with the different atoms in the molecule. The overall final state may differ e.g. by mirroring the projectile velocity vector.

A first, "simple" experiment is proposed here, where the final ionic states A* and C* are compared, when the projectile is impinging from "left" or from "right". Reducing the ion velocity far below the speed of light, the time delay range can be extended. This kind of measurement could already now provide new inside into the range of zeptosecond electronic dynamics. The method proposed here to use ions moving with the speed of light as ultra-fast clocks allows the investigation of fundamental features of "Locality" or "Non-Locality" in quantum systems, i.e., whether the information exchange occurs in such systems instantaneous or only by speed of light.

Theorists may be convinced that the questions raised here are already answered. But nevertheless one should experimentally verify any fundamental theoretical prediction. When Otto Stern decided in 1920 to perform the "Stern-Gerlach-Experiment" [19] and later in 1933 to measure the magnetic moment of the proton [20] theorists tried to convince him, such difficult experiments should not be done since theory had already answered these questions. Nevertheless, he performed his milestone measurements and could disprove theory.

4 Conclusion

This paper does show that since more than 100 years the motion of atoms, molecules or ions was successfully used to measure dynamical features (like lifetimes, phase shifts etc.) in quantum systems. Already more than 30 years ago quantum beat structures with 10 zeptoseconds resolution could be measured about factor 100 to 1000 shorter than present pump & probe two-pulse laser techniques can achieve. The future heavy ion facility FAIR [21] will provide relativistic heavy ion beams with which the here proposed ion-atom/molecule pump & probe technique can be performed to explore dynamics in the zeptosecond regime.

Acknowledgements We thank Siegbert Hagmann, Hans Jürgen Lüdde, John Briggs, Mike Prior, and C. L. Cocke for many helpful discussions. Furthermore, we are indebted to the Deutsche Forschungsgemeinschaft and the BMFT for financial support as well as Roentdek company for technical support in performing the experiments.

References

1. W. Kühlbrandt, The resolution revolution. Science, **343**, 1443 (2014) and other references therein
2. F. Calegari, G. Sansone, M. Nisoli, *Attosecond Pulses for Atomic and Molecular Physics* (2014) Book Lasers in Materials Science, pp. 125–141, Springer International Publishing; A. L. Cavaleri et al., Nature (London) 449, 1029 (2007)

3. R. Dörner, V. Mergel, O. Jagutzki, L. Spielberger, J. Ullrich, R. Moshammer, H. Schmidt-Böcking, Cold target recoil ion momentum spectroscopy: a 'momentum microscope' to view atomic collision dynamics. Phys. Rep. **330**, 95–192 (2000); R. Dörner, T. Weber, M. Achler, V. Mergel, L. Spielberger, O. Jagutzki, F. Afaneh, M.H. Prior, C.L. Cocke, H. Schmidt-Böcking, 3-D coincident imaging spectroscopy for ions and electrons, imaging in chemical dynamics, in *ACS Symposium Series*, vol. 770, ed. by A. Suits, R. E. Continetti, (Oxford Univ. Press, 2001), pp. 339–349; R. Dörner, H. Schmidt-Böcking, V. Mergel, Th. Weber, L. Spielberger, O. Jagutzki, A. Knapp, H.P. Bräuning, From atoms to molecules. *Many Part. Quantitative Dynamic Atomic Molecular Fragm.*, ed. by V.P. Shevelko, J. Ullrich (Springer Verlag, 2002); J. Ullrich, R. Moshammer, A. Dorn, R. Dörner, L.Ph.H. Schmidt, H. Schmidt-Böcking Recoil-ion and electron momentum spectroscopy: reaction-microscopes. Rep. Prog. Phys. **66** 1463–1545 (2003)
4. M. Born, R. Oppenheimer, Zur Quantentheorie der Molekeln. Ann. Phys. **389**(20), 457–484 (1927)
5. A.H. Zewail, Femtochemistry: recent progress in studies of dynamics and control of reactions and their transition states. J. Phys. Chem. (Centennial Issue) **100**, 12701 (1996); A.H. Zewail, Femtochemistry: atomic-scale dynamics of the chemical bond†. J. Phys. Chem. A. **104**(24), 5660–5694 (2000), adapted from the Nobel Lecture; F. Krausz and M. Ivanov, 2009 Rev. Mod. Phys. **81**, 163; M. Drescher, M. Hentschel, R. Kienberger, G. C. Tempea, C. Spielmann, G.A. Reider, P.B. Corkum, F. Krausz, Science **291**, 1923 (2001)
6. L.Ph.H. Schmidt, S. Schössler, F. Afaneh, M. Schöffler, K. Stiebing, H. Schmidt-Böcking, R. Dörner, Young-type interference in collisions between hydrogen molecular ions and helium. Phys. Rev. Lett. **101**, 173202 (2008)
7. T. Jahnke, L. Cederbaum T. Jahnke, A. Czasch, M. S. Schöffler, S. Schössler, A. Knapp, M. Käsz, J. Titze, C. Wimmer, K. Kreidi, R. E. Grisenti, A. Staudte, O. Jagutzki, U. Hergenhahn, H. Schmidt-Böcking, R. Dörner, Experimental observation of interatomic coulombic decay in neon dimers. Phys. Rev. Lett. **93**, 163401 (2004); T. Jahnke, A. Czasch, M. Schöffler, S. Schössler, M. Käsz, J. Titze, K. Kreidi, R.E. Grisenti, A. Staudte, O. Jagutzki, L.Ph.H. Schmidt, Th. Weber, H. Schmidt-Böcking, K. Ueda, R. Dörner, Experimental separation of virtual photon exchange and electron transfer in interatomic coulombic decay of neon dimers. Phys. Rev. Lett. **99**, 153401 (2007); L.S. Cederbaum, J. Zobeley, and F. Tarantelli; Giant intermolecular decay and fragmentation of clusters. Phys. Rev. Lett. **79**, 4778 (1997)
8. J. Burgdörfer, C. Lemmel, X. Tong, Invited Lecture at ICPEAC 2019, arXiv:2001.02900v1 [quant-ph] 9 Jan 2020
9. R. Schuch, M. Meron, B.M. Johnson, K.W. Jones, R. Hoffmann, H. Schmidt-Böcking, I. Tserruya, Quasimolecular X-ray spectroscopy for slow Cl^{16+}-Ar collisions. Phys. Rev. A **37**, 3313 (1988)
10. I. Tserruya, R. Schuch, H. Schmidt-Böcking, J. Barrette, Wang Da-Hai, B.M. Johnson, M. Meron, K.W. Jones, Interference effects in the quasimolecular K X-ray production probability for 10 MeV Cl^{16+}- Ar collisions. Phys. Rev. Lett. **50**, 30 (1983); R. Schuch, H. Schmidt-Böcking, I. Tserruya, B.M. Johnson, K.W. Jones, M. Meron, X-ray spectroscopy of Cl-Ar molecular orbitals from $1s\sigma$-$2p\pi$ transitions. Z. Phys. A **320**, 185 (1985)
11. R. Schuch, H. Ingwersen, E. Justiniano, H. Schmidt-Böcking, M. Schulz, F. Ziegler, Interference effects in K-vacancy transfer of hydrogen like S Ions with Ar. J. Phys. B **17**, 2319 (1984); R. Schuch, H. Ingwersen, E. Justiniano, H. Schmidt-Böcking, M. Schulz, F. Ziegler, Experiments with decelerated S^{15+}-beams: interferences in K-shell to K-shell charge transfer, atomic and nuclear heavy ion interactions. Centr. Inst. of Phys. (1986); S. Hagmann, J. Ullrich, S. Kelbch, H. Schmidt-Böcking, C.L. Cocke, P. Richard, R. Schuch, A. Skutlartz, B. Johnson, M. Meron, K. Jones, D. Trautmann, F. Rösel, K-K-charge transfer and electron emission for 0.13 MeV/u F^{8+} + Ne collisions. Phys. Rev. A **36**, 2603 (1987); R. Schuch, M. Schulz, Y.S. Kozhedub, V.M. Shabaev, I.I. Tupitsyn, G. Plunien, P. H. Mokler, H. Schmidt-Böcking, Quantum Interference of K Capture in Energetic Ge^{31+}(1s)-Kr Collisions
12. J.M. Dahlström, A.I. Huillier, A. Maquet Introduction to attosecond delays in photoionization. J. Phys. B. At. Mol. Opt. Phys. **45**, 183001 (2012); J.M. Dahlström and E. Lindroth, Study

of attosecond delays using perturbation diagrams and exterior complex scaling, J. Phys. B: Atomic Mol. Opt. Phys. **47**, 124012 (2014)

13. O. Stern, M. Volmer, Über die Abklingungszeit der Fluoreszenz. Physik. Z. **20**, 183–188 (1919)
14. S. Bashkin, Nucl. Instrum. Methods **28**: 88 (1964); S. Bashkin, ed., *Beam-Foil Spectroscopy* (Gordon and Breach, New York 1968); S. Bashkin and I. Martinson, J. Opt. Soc. Am. **61**: 1686 (1971); S. Bashkin (ed.), Beam-Foil Spectroscopy Springer, Berlin (1976); H G Berry 1977 Rep. Prog. Phys. **40**, 155. View the article online for updates and enhancements. Related content
15. N.F. Ramsey, A molecular beam resonance method with separated oscillating fields. Phys. Rev. **78**, 695 (1950); N. Ramsey, Rev. of Mod. Phys. **62**(3) 541–552 (1990)
16. W. Iskandar, J. Matsumoto, A. Leredde, X. Fléchard, B. Gervais, S. Guillous, Atomic site-sensitive processes in low energy ion-dimer collisions. Phys Rev Lett. **113**, 14, 143201
17. R. Moshammer, J. Ullrich, H. Kollmus, W. Schmitt, M. Unverzagt, H. Schmidt-Böcking, R.E. Olson, The dynamics of target single and double ionization induced by the virtual photon field of fast heavy ions, X-ray and inner-shell processes. AIP Conf. Proc. **389**, 153 (1996); R. Moshammer, W. Schmitt, J. Ullrich, H. Kollmus, A. Cassimi, R. Dörner, O. Jagutzki, R. Mann, R.E. Olson, H.T. Prinz, H. Schmidt- Böcking, L. Spielberger, Ionization of helium in the attosecond equivalent light pulse of 1 GeV/nucleon U^{92+} projectiles. Phys. Rev. Lett. **79**, 3621 (1997); R. Moshammer, J. Ullrich, W. Schmitt, H. Kollmus, A. Cassimi, R. Dörner, R. Dreizler, O. Jagutzki, S. Keller, H.J. Lüdde, R. Mann, V. Mergel, R.E. Olson, Th. Prinz, H. Schmidt-Böcking, L. Spielberger, Photodisintegration of atoms in the attosecond equivalent light pulse of highly charged relativistic ions. Phys. Rev. Lett. **79**, 3621 (1997); C.F. von Weizsäcker, Z. Phys. **88**, 612 (1934); E. J. Williams, Kgl. Danske Videnskab. Selskab Mat.-fys. Medd. **13**(4) (1935)
18. K. Fehre, D. Trojanowskaja, J. Gatzke, M. Kunitski, F. Trinter, S. Zeller, LPhH Schmidt, J. Stohner, R. Berger, A. Czasch, O. Jagutzki, T. Jahnke, R. Dörner, M.S. Schöffler, Absolute ion detection efficiencies of microchannel plates and funnel microchannel plates for multi-coincidence detection. Rev. Sci. Instrum. **89**, 045112 (2018)
19. O. Stern, Ein Weg zur experimentellen Prüfung der Richtungsquantelung im Magnetfeld. Z. Physik **7**, 249–253 (1921)
20. ETH-Bibliothek Zürich, Archive, http://www.sr.ethbib.ethz.ch/, Otto Stern tape-recording Folder» ST-Misc.«, 1961 at E.T.H. Zürich by Res Jost
21. https://www.gsi.de/forschungbeschleuniger/fair.htm

5

Otto Stern's Legacy in Quantum Optics: Matter Waves and Deflectometry

Stefan Gerlich, Yaakov Y. Fein, Armin Shayeghi, Valentin Köhler, Marcel Mayor and Markus Arndt

Abstract Otto Stern became famous for molecular beam physics, matter-wave research and the discovery of the electron spin, with his work guiding several generations of physicists and chemists. Here we discuss how his legacy has inspired the realization of universal interferometers, which prepare matter waves from atomic, molecular, cluster or eventually nanoparticle beams. Such universal interferometers have proven to be sensitive tools for quantum-assisted force measurements, building on Stern's pioneering work on electric and magnetic deflectometry. The controlled shift and dephasing of interference fringes by external electric, magnetic or optical fields have been used to determine internal properties of a vast class of particles in a unified experimental framework.

1 From Otto Stern to Universal Molecule Interferometry

Our contribution honors the legacy of Otto Stern, who paved the path for 100 years of exciting research into atomic and molecular beams, spin physics and matter-wave interferometry. Many of his ideas and original methods are still implemented in our present-day experiments. On the one hand it is impressive how much progress these fields have made since Stern's time, but it is also humbling to realize how many of Stern's experimental challenges remain even in the most advanced experiments today.

S. Gerlich · Y. Y. Fein · A. Shayeghi · M. Arndt (✉)
Faculty of Physics, University of Vienna, Boltzmanngasse 5, 1090 Vienna, Austria
e-mail: markus.arndt@univie.ac.at

V. Köhler · M. Mayor
University of Basel, St. Johannsring 19, 4056 Basel, Switzerland
e-mail: marcel.mayor@unibas.ch

1.1 Stern's Legacy in Molecular Beam Deflection

It is enlightening to look at one of Stern's early papers, *Zur Methode der Molekularstrahlen* [1], in which he describes the first applications of atomic and molecular beam deflectometry and formulates criteria for achieving the highest possible metrological sensitivity. His idea was straightforward and is sketched in Fig. 1a: a beam of atoms or molecules is launched into high vacuum, collimated, deflected by external fields and detected downstream with position resolution. Following Stern's notation, the deflection s of a particle of mass m after traveling a distance l with a velocity v in a uniform force field K is

$$s = \frac{1}{2}\frac{K}{m}\frac{l^2}{v^2}. \quad (1)$$

Knowing the beam velocity, geometry and fields involved, it is then straightforward to extract electronic or magnetic properties of the atoms or molecules, since they are contained within K. Stern formulated three criteria to achieve high sensitivity for deflectometry[1]:

1. **Narrow beam width**: *"make the beam as narrow as possible, because the narrower it is, the smaller the deflection* s *we can measure"*.
2. **Large deflection region**: *"make the path* l *through the field as long as possible, because* $s \sim l^2$*"*.
3. **Strong fields**: *"make the force* K *as big as possible, because* $s \sim K$*"*.

Since the first two criteria reduce the flux of detected particles, Stern proposed to "*...increase the intensity to the required amount by placing 100 identical furrows next to each other on the pole shoe, all of them pointing to the same detector area, such that their images fall on top of each other.*"

Stern envisioned far-reaching applications of his beam deflection apparatus, such as the measurement of nuclear magnetic moments,[2] induced moments, electric dipole moments, and higher-order moments. Many of our experiments with atoms, molecules and clusters are rooted in these ideas.

1.2 Stern's Legacy in Matter-Wave Research

Our experiments are also based on a second series of pioneering studies by Otto Stern: while textbooks correctly ascribe the first demonstration of matter-wave diffraction to the electron experiments by Davisson and Germer [3], it is noteworthy that Estermann and Stern were already working toward matter-wave experiments in 1926. In 1930

[1]Since Otto Stern's early papers were written in German, we use our own translation where a verbatim citation is indicated.

[2]Stern's Nobel Prize in 1943 was awarded for his measurement of the proton's magnetic moment.

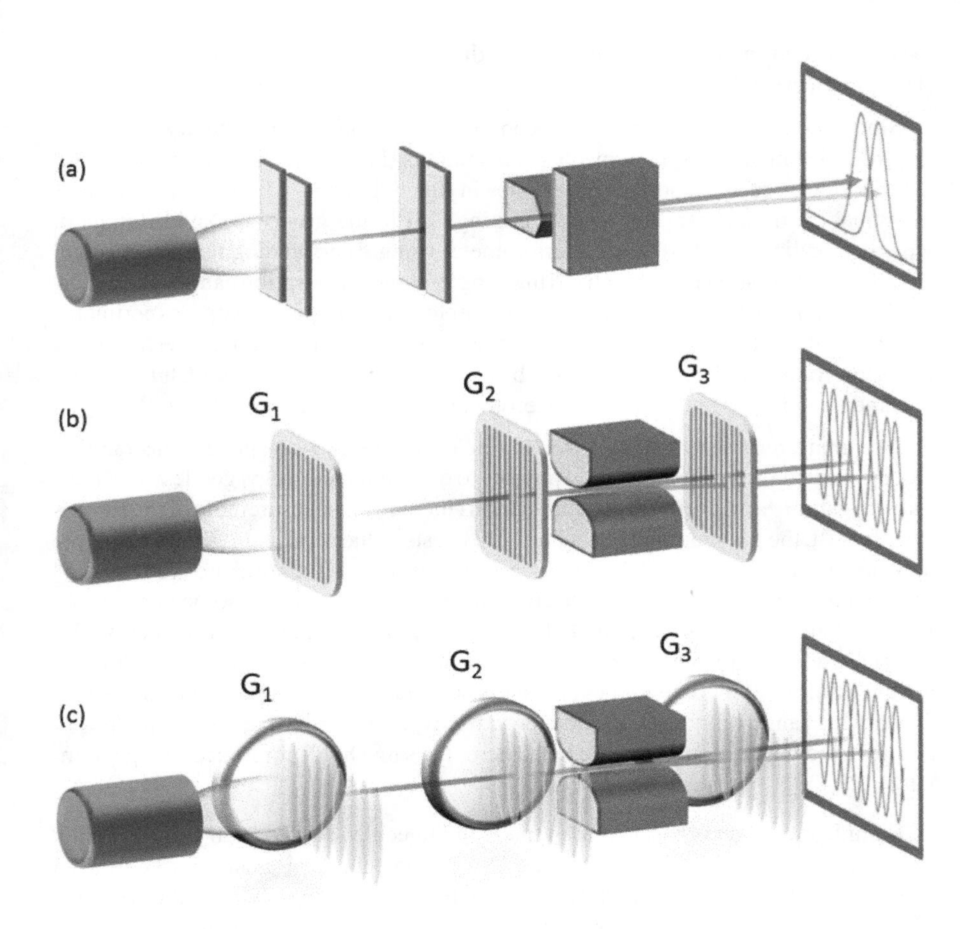

Fig. 1 **a** A molecular beam deflection experiment in the spirit of Otto Stern consists of an intense beam source, narrow collimators, an inhomogeneous deflection field and a position resolving detector. **b** Deflectometry in a near-field Talbot-Lau interferometer (TLI): the first grating, G_1, prepares transverse coherence from an initially incoherent beam, while the second grating imparts a superposition of momenta to the delocalized matter wave. Interference manifests itself as a particle density pattern with the same periodicity of as the gratings which can be detected by scanning the third grating and counting the transmitted particles. **c** Near-field interferometry requires either coherent sources or an absorptive first grating G_1. This can be realized by a material mask or photo-induced depletion of the molecular beam in a standing light wave grating [2]. In Sec. 2, different variants of Talbot-Lau interferometers are discussed, including the all-optical OTIMA experiment and the Kapitza-Dirac Talbot-Lau scheme, which constitutes a hybrid of (**b**) and (**c**) with G_1 and G_3 as material gratings, but an optical phase grating serving as G_2

they succeeded in demonstrating the first diffraction of atoms (He) and molecules (H_2) from a crystal surface [4].

Since then, huge progress has been made in atom interferometry based on improved beam sources, nanomechanical gratings, laser physics and optical beam-splitting techniques. Important milestones in this field are the first atom diffraction at optical [5, 6] and nanomechanical gratings [7] in the group of David Pritchard, who also realized the first atom interferometer using free-standing nanomechanical gratings [8]. Christian Bordé realized that single-photon absorption can act as a coherent beam splitter for atoms [9] and reinterpreted earlier spectroscopy experiments on SF_6 [10] as Ramsey-Bordé interferometry. A time-domain atom interferometer based on Raman transitions was built by Mark Kasevich and Steven Chu [11] and became a model for many atom interferometer realizations around the world.

Atom interferometry is now a thriving field of physics with applications ranging from precision tests of fundamental physics to quantum metrology, geodesy and inertial navigation. Some noteworthy applications include the measurement of the Earth's gravity [12], the gravitational constant G [13], tests of the weak equivalence principle and the universality of free-fall [14], measurements of the fine structure constant [15] and rotation sensing [16, 17]. Interferometry experiments have also been proposed for gravitational wave detection [18] and for dark matter [19] and dark energy [20, 21] searches. For reviews covering these topics see e.g. Refs. [22–24]. With the advent of ultra-cold quantum degenerate gases and Bose-Einstein condensates [25, 26], a wide range of mesoscopic matter-wave experiments have also become possible, with too many examples to be listed here; the same holds for molecular quantum gases [27, 28].

Significant progress has also been made in molecule interferometry since Stern's early experiments with H_2. Diffraction at a nanomechanical mask was key to the discovery of the extremely weakly bound helium dimer He_2 [29] and the basis for Mach-Zehnder interferometry with Na_2 [30]. The combination of four $\pi/2$-pulse beam splitters in a Ramsey-Bordé interferometer was demonstrated with I_2 [31]. We refer to contributions by Jan-Peter Toennies, Wieland Schoellkopf and David Pritchard in this book for more on these topics.

Stern's original idea was to exploit beam deflectometry as a tool to learn about the inner structure and physics of atoms and molecules, and the techniques of atom interferometry have enabled a number of such measurements. The nature of the bonding of He_2 [32] was measured via quantum reflection from a grating, since other techniques would have been too invasive to probe the extremely fragile bond. The static polarizability of sodium [33], lithium [34] and other alkali atoms [35] was measured using atom interferometry, and long-range potential properties [36], van der Waals coefficients [37], atomic tune-out wavelengths [38] and transition matrix elements [39, 40] as well as surface excitations [41] have all been studied using matter-wave diffraction.

2 Interferometer Concepts for Studying the 'Wave-Nature of Everything'

Building on the work with atoms and dimers and fueled by advances in lasers and nanotechnology, the investigation of the quantum nature of more massive objects became possible by the end of the 20th century.

The fullerene C_{60} was the first complex molecule for which de Broglie interference was demonstrated in **far-field diffraction** [42]. Fullerenes are particularly well-suited for beam experiments since they are thermally stable and can be evaporated in a simple furnace. Since a thermal source lacks both transverse and longitudinal coherence, the particles were sent through a pair of collimation slits to generate the required transverse coherence before being diffracted at a nanofabricated grating with a period of $d = 100\,\text{nm}$. The far-field diffraction pattern is a convolution of the single-slit and multi-slit pattern as familiar from optics textbooks, with the relevant wavelength in this case $\lambda_{dB} = h/mv$. The molecular density pattern was detected by scanning a tightly focused green laser over the molecular beam to cause thermal ionization and create countable ions [43]. The experiment is illustrated in Fig. 2.

At first glance it is intriguing that high-contrast interference could be observed despite the fact that the molecules were heated to about 900 K, thus exciting many rotational and vibrational levels. While each individual molecule is distinguishable by virtue of its unique internal state, interference is still observed because each particle interferes with itself, and the evolution of its vibrational and rotational modes occurs simultaneously along each arm of the interferometer. The center-of-mass motion is the only relevant degree of freedom—as long as coupling to the environment can be suppressed.

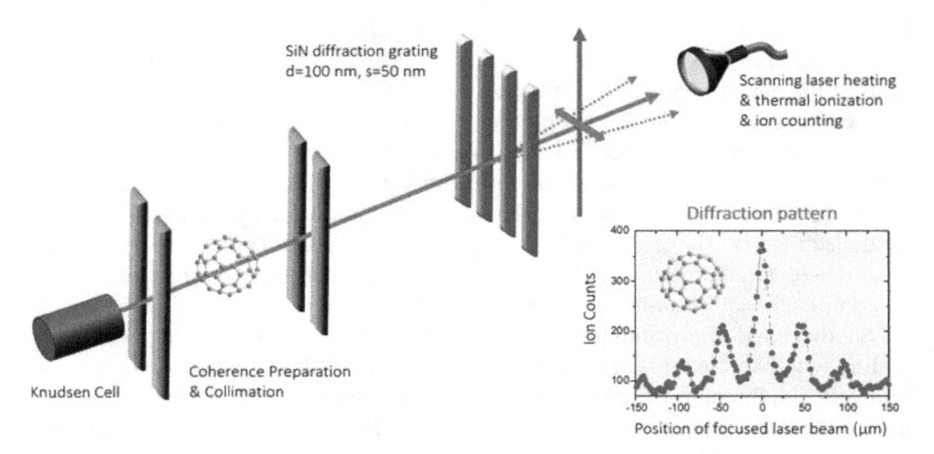

Fig. 2 Fullerene diffraction at a 100 nm period nanomechanical grating as realized in Vienna [42, 44]. For a central velocity of 136 m/s, the de Broglie wavelength is 4 pm and thus less than 200 times the molecular diameter. Two slits of 10 μm width separated by 104 cm prepare the transverse coherence required to illuminate several slits of the grating

A more visual way of revealing the molecular wave-particle duality is via fluorescence imaging. Starting from a micron sized laser emission source, molecules were diffracted at a nanomechanical mask and deposited on a quartz slide, where they were detected in real time by fluorescence microscopy with a spatial resolution of about 10 nm [45].

The de Broglie wavelength of a particle with $m \simeq 1000$ Da travelling at 100–300 m/s is of order $\lambda_{\mathrm{dB}} \simeq 10^{-12}$ m. Typical gratings – both nanomechanical masks and standing light waves – have periods of $d \simeq 100$ nm or larger, which yield diffraction angles of 10 μrad. Resolving this small angle requires collimation and an angular resolution of the detector of a few μrad. Increasing the mass by a factor of ten thus requires improving the collimation and detector resolution by the same factor.

Textbook-style far-field diffraction thus becomes quickly impractical for beams of high-mass particles. Near-field optics, however, provides a viable solution. The phenomenon of lens-less grating self-imaging was first observed with light by Henry Fox Talbot in 1836 [46] and later extended by Ernst Lau to incoherent sources [47]. A **Talbot-Lau interferometer** (TLI) relies on this self-imaging phenomenon and can be formed with two gratings, as illustrated in Fig. 1b. In the symmetric configuration, two identical gratings are spaced apart by a multiple of the Talbot length, $L_T = d^2/\lambda_{\mathrm{dB}}$, with d the grating period. In general, near field diffraction causes a complicated pattern, known as a Talbot carpet, to be imprinted into the light or matter-wave beam behind the second grating. At certain distances (in the symmetric case simply the same distance as the G_1–G_2 separation), the pattern is an exact self-image of the second grating, i.e., fringes with period d. A third grating with the same period can be employed to detect this self image: scanning it transversely to the beam will yield a sinusoidal modulation in the flux as detected by a spatially integrating detector.

The TLI scheme has several advantages for interferometry of massive particles. Compared to far-field schemes like the Mach-Zehnder interferometer, a TLI has relaxed requirements on transverse coherence, permitting the use of relatively uncollimated molecular beams. This is because each opening of the first grating acts as a coherent source for the second grating, with the symmetry of the setup ensuring that the many interferometer trajectories emanating from the first grating recombine in phase at the position of the third grating. In other words, the Talbot-Lau scheme benefits largely from multiplexing, a strategy already envisioned by Stern for deflectometry. The large gain in throughput is an important asset especially when dealing with large (organic) molecules, which are typically very fragile and difficult to volatilize intact, resulting in low beam intensities.

Another reason for working with near-field interferometers is due to their favorable scaling with particle mass, as pointed out by Clauser [48]. For a given length of the setup, the minimum resolvable de Broglie wavelength scales with d^2, in contrast to d in far-field diffraction. Increasing m by a factor of 100 – thereby reducing λ_{dB} by the same factor – would require gratings with a 10 times smaller period d in a near-field scheme, whereas they would have to be 100 times smaller to observe the diffraction of the same mass in the far-field. Talbot-Lau interferometry thus allows us to access a high mass range even with moderate interferometer baselines and grating periods.

There are a variety of ways to realize gratings in the lab, with the restriction that G_1 and G_3 must be transmission masks to fulfill their roles of spatially confining the beam and spatially filtering the beam, respectively.

Nanofabricated masks can be used for a large variety of particles because their action does not depend on any specific internal particle property or transition. Fullerenes once again served as the first species to be studied in a three-grating TLI[3] at the University of Vienna [49]. A TLI setup was also used to observe the wave nature of tetraphenylporphyrins (TPP) and the fluorofullerenes $C_{60}F_{48}$ [50] as well as to demonstrate the prospects of molecule lithography [51].

Material gratings, however, also induce strongly velocity-dependent (dispersive) Casimir-Polder phase shifts on matter waves, which limits the maximal fringe contrast particularly for highly polarizible, slow molecules [52]. This particle-wall interaction may also be enhanced by local charges deposited in the fabrication process. Surface effects can be partially mitigated by reducing the grating thickness, but even at the ultimate limit of an atomically thin diffraction grating made from single-layer graphene [53], a sizeable phase shift remains.

Optical beam splitters are particularly appealing for molecule interferometry, since they are free from effects of surface geometry and quality, contamination and charges, and they can be precisely defined both spatially and temporally. Diffraction at a standing light wave was originally proposed by Kapitza and Dirac for electrons in the Bragg regime [54]. It was first realized with atoms using off-resonant optical dipole force phase gratings in the Raman-Nath regime [5].[4] Kapitza-Dirac diffraction was demonstrated both for electrons [55] and fullerenes [56] in 2001.

While the polarizability of atoms varies by several orders of magnitude around an optical resonance, it does not typically vary by more than 50% across a large part of the optical spectrum for complex, warm molecules or nanoparticles. An optical phase grating of a fixed wavelength may therefore serve as a universal coherent beam splitter for a large variety of particles. This universality, however, comes at a price: without resonant enhancement, the polarizability remains moderate and the process requires high laser intensities. The wavelength should also be chosen to avoid photon absorption and re-emission processes, which tend to reduce interference contrast.

The benefits of both optical gratings and of near-field interferometry led to the proposal of a Talbot-Lau interferometer whose central element G_2 is an optical phase grating (see Fig. 1) [57]. In this scheme, the **Kapitza-Dirac-Talbot-Lau Interferometer** (KDTLI), the first and last grating remain material masks while the central grating is formed by a thin standing light wave with a period matching that of the outer two gratings. Such an interferometer was built and successfully employed with a variety of complex molecules with masses up to 10^4 Da [52, 58, 59].

[3]From here on, TLI will refer to Talbot-Lau interferometers implemented with three nanomechanical gratings.

[4]Modern matter-wave literature often associates the names of Kapitza and Dirac with diffraction at a thin dipole force phase grating, even though their original idea applied to non-polarizable electron diffracted at the ponderomotive potential created by interaction with the light field.

The use of nanomechanical gratings for G_1 and G_3 will eventually be limited when the Casimir Polder potential becomes sufficiently strong (for slow, highly-polarizable molecules) that the particles are completely deflected to the grating walls and cannot pass. It was therefore proposed to form transmission gratings using a standing light wave. This can work via photo-ionization, where particles passing the anti-nodes of a standing light wave are ionized and removed from the beam, leaving every node as an effective grating slit [2]. This process works best if the photon energy exceeds the ionization energy and the absorption cross section is sufficiently high to allow absorption of more than one photon in every anti-node, conditions which can be met, for example, by tryptophan-rich peptides or low work function metal clusters.

The **Optical TIme-domain MAtter-wave interferometer** (OTIMA) is a time-domain interferometer that employs pulsed absorptive optical gratings [60, 61]. Since all particles from the pulsed source interact with the same grating pulse at the same time, independent of their position, various dispersive phase shifts are eliminated, most prominently the shift related to Earth's gravitational acceleration: $\Delta\varphi \propto kgT^2$, where $k = 4\pi/\lambda_L$ is the wave number of the optical grating and T the pulse separation time. However, phase shifts with explicit velocity dependence remain, such as the Coriolis shift induced by the rotation of the Earth, $\Delta\varphi \propto k(2\vec{v} \times \vec{\Omega}_E)T^2$.

Photo-depletion gratings have been extensively studied in atom interferometry [62, 63]. For molecular quantum optics, we are developing mechanisms that can act on the widest class of particles possible. The fluorine (F_2) grating lasers in the OTIMA experiment have a wavelength of $\lambda_L = 157.6\,\text{nm}$ (7.9 eV). This suffices to ionize van der Waals clusters of aromatic anthracene and caffeine [61, 64] as well as tryptophan-rich polypeptides [65]. Combined with ultrafast desorption techniques [65], the first interference of a natural antibiotic polypeptide, Gramicidin A1, was recently demonstrated in this experiment [66] (see Fig. 3a).

Photo-fragmentation is another mechanism which can be used to achieve an optical transmission grating, as demonstrated with van der Waals clusters of hexafluorobenzene [64]. A major goal is to develop single-photon cleavage of tagged biopolymers for future interference experiments with proteins and DNA. The cleavage mechanism itself has recently been successfully demonstrated with functionalized insulin [68].

Since the grating periods of KDTLI and OTIMA are already at the lower limit of commercially available high power lasers, pushing high mass quantum experiments even further requires increasing the flight time between the gratings. This may be achieved by advanced particle cooling schemes or by increasing the interferometer length. The **Long-baseline Universal Matter-wave Interferometer** (LUMI) is a ten-fold stretched realization of the KDTLI experiment. This instrument accepts de Broglie wavelengths as small as 35 fm and has already demonstrated interference with a molecular library centered at 27,000 Da, with a typical molecule in the library consisting of nearly 2000 atoms and travelling at 260 m/s [67] (see Fig. 3b). These molecules were synthesized at the University of Basel for the purpose of these interference experiments, as described in Sect. 3.2. To date, they are the most massive and complex objects for which quantum interference has been demonstrated.

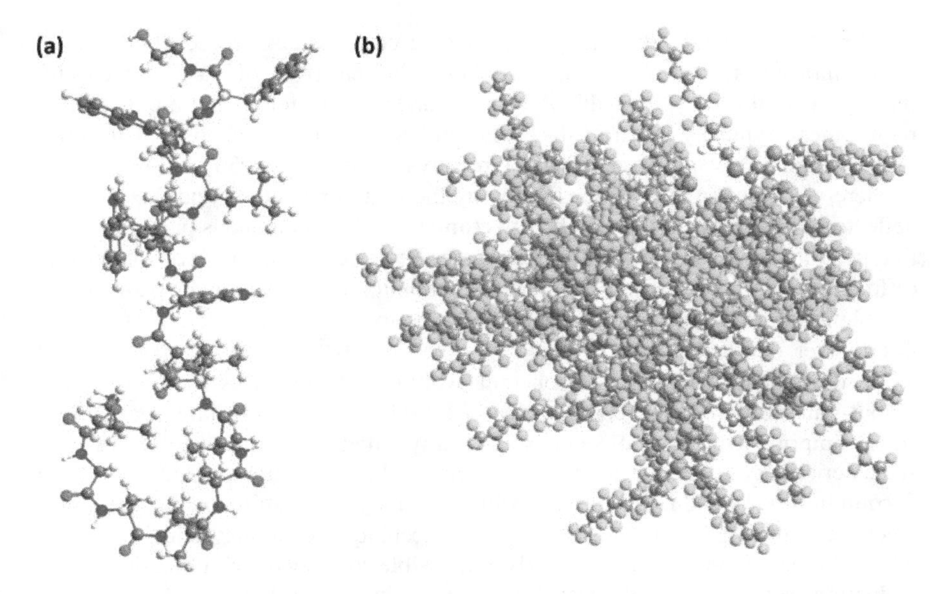

Fig. 3 **a** The antibiotic pentadecapeptide Gramicidin A1 is the most complex natural biomolecule in matter-wave experiments so far [66]. **b** A library of perfluoroalkyl-functionalized oligoporphyrins sets the current record for the most massive particles seen in matter-wave interference to date [67]

The LUMI design is modular: the central grating can be interchanged between a nanomechanical grating and an optical phase grating. Generalized versions of the scheme, including optical transmission gratings and surface detection have also been theoretically investigated [69]. The present LUMI experiment is compatible with both atoms and complex molecules, and an upgrade that is currently being implemented will allow it to also work with massive metal clusters. In this sense LUMI is a truly 'universal' interferometer as well as a powerful instrument for metrological studies.

3 Quantum-Assisted Deflectometry of Atoms and Complex Molecules

Otto Stern demonstrated that molecular beam methods can boost the precision in the measurement of atomic and molecular properties, external forces or fundamental constants. The Stern-Gerlach experiment has become standard textbook material, and his electric deflection experiments paved the way for a wide range of studies in physics and chemistry in the decades that followed.

Classical beam deflectometry measures the shift of a tightly collimated molecular beam as a result of its interaction with an external field (see Fig. 1 and Eq. (1)) [70–74]. The magnitude of the shift encodes information about the field and the molecular coupling to it. Beam deflectometry can also be used to separate isomers [72, 75, 76] or sort molecules by their quantum state [77]. The sensitivity of classical measurements is determined by the width of the beam and the detector resolution. Beam flux requirements typically constrain the beam width to $> 10\,\mu m$ while position-sensitive time-of-flight mass spectrometers can typically resolve shifts $> 50\,\mu m$ [78].

Here we review a method for improving the spatial resolution and flux in beam deflectometry: **Quantum-assisted deflectometry**. The technique is particularly fitting in a tribute to Otto Stern since it combines beam deflectometry and matter-wave diffraction, two fields he helped pioneer. Combining these techniques gains orders of magnitude in spatial resolution, allowing us to resolve nanometer, rather than micrometer, deflections.

As described in Sect. 2, the coherent evolution of molecules in a generalized Talbot-Lau interferometer (e.g. TLI, KDTLI, OTIMA, LUMI) manifests as density fringes imprinted into the molecular beam. In a symmetric setup these fringes have the same periodicity as the interferometer gratings, with $d = 79\,\text{nm}$ in OTIMA and $d = 266\,\text{nm}$ in KDTLI and LUMI. The capability to track fringe shifts on the nanometer level yields the high spatial resolution of this technique, enabling the measurement of tiny forces which would be nearly impossible to resolve with classical beam deflection methods. In analogy to Eq. (1), the fringe shift due to a uniform force acting transversely along the entire interferometer is

$$s = k\frac{K}{m}T^2 = k\frac{K}{m}\frac{L^2}{v^2} \tag{2}$$

where $k = 2\pi/d$, T the time between gratings (for pulsed experiments like OTIMA) and L the inter-grating separation.

3.1 *Quantum versus Classical Deflection*

In the matter-wave deflectometry experiments described here, a force is applied to the molecules and the resulting phase shift of the interference fringes is detected. Consider a particle beam in a uniform electric field gradient. It will be broadened and/or shifted depending on whether the particles have a permanent electric dipole moment or are only polarizible. In quantum-assisted deflectometry this corresponds to contrast reduction and the deflection of the fringe pattern, where the fringes provide a ruler with high spatial resolution.

The Talbot-Lau deflectometry scheme can be compared to similar experiments using Mach-Zehnder interferometry [33] and classical Moiré deflectometry [79]. The three schemes are limiting cases of the same physical setup: a molecular beam traversing three gratings of period d each separated by a distance L. The relevant length scales are the Talbot length $L_T = d^2/\lambda_{dB}$ and the "aperture Talbot Length", $L_a = a^2/\lambda_{dB}$, where a is the width of the beam at the first grating.

1. The *far-field (Mach-Zehnder)* regime is reached when the grating separation satisfies $L \gg L_a$. In this limit, the diffraction orders emerge from G_1 as distinct partial beams which are diffracted back by G_2 and recombined by G_3. The isolated partial beams can be made to locally interact with potentials, which induce a measurable phase shift in the interference pattern. This setting was realized in the first atom interference [8] and metrology [33] experiments.
2. If the grating separation satisfies $L_a/N > L > L_T$, with N the number of illuminated grating slits, the apparatus realizes a *Talbot-Lau interferometer* [80, 81], which is the regime of our molecule interference experiments. Each individual molecule is still spatially delocalized across several, or in some cases up to 100, grating periods [82]. However, the molecular beam is so wide and the diffraction angles so small that all of these partial interferometers overlap. While this makes it impossible to address individual partial beams, the symmetry of the interferometer ensures that all partial waves converge at the position of the third grating, giving rise to high-contrast interference fringes. A potential gradient can be applied within the interferometer to induce an envelope phase shift of the interference fringes.
3. The third limit is that of classical Moiré deflectometry, in which $L_a, L_T \gg L$. In this setting, the wavelets originating in G_1 do not evolve fast enough to cover two slits in G_2 coherently. Even in this classical regime, sensitive force sensing is still possible [79]. This is the closest realization to the classical beam multiplexing proposed by Stern.

Among the three different regimes, only the far-field Mach-Zehnder features well-separated interferometer arms, and it still holds the record for the most sensitive polarizability measurements [33–35]. In this regime one can also explore topological and geometric phases, such as the Aharanov-Bohm [83], Aharanov-Casher [84], Berry [85] and He-McKellar-Wilkens phases [86].

The TLI regime, on the other hand, has the best mass scalability of the three limits and it is therefore currently the only setting compatible with quantum-enhanced measurements of large molecules. In both the far- and near-field limits (1 and 2 above), the sensitivity to external forces depends on the enclosed interferometer area and the detected signal-to-noise ratio. Compared to a classical Moiré deflectometer, the TLI employs smaller grating periods and/or longer machine length and therefore has intrinsically better sensitivity to small fringe displacements.

3.2 Molecules for Interferometry: Choice, Synthesis and Sources

3.2.1 General Strategies

Among all nanoscale particles, molecules are ideally suited for matter-wave experiments due to their monodisperse nature. Being virtually identical, they also exhibit a very narrow isotopomeric mass distribution, typically within a few Daltons. Over the years, we have explored a large variety of structures, from commercially available molecules to tailor-made model compounds with properties optimized for the particular experiment. The collaboration between experimental physicists and synthetic chemists has enabled access to higher mass regimes and the development of new diffraction mechanisms.

Three challenges need to be considered when selecting molecules for interference experiments: the preparation of neutral particle beams, novel diffraction mechanisms and detection schemes with high sensitivity and resolution. For different molecules different techniques may apply and one research goal is to find the most generic combinations that allow treating the largest class of particles.

While thermal evaporation from a Knudsen cell is a simple experimental technique, it requires molecules with sublimation temperatures below their degradation temperature to guarantee the launch of individual and intact molecules of known composition and mass. This calls for a molecular design of thermally stable molecules with minimal intermolecular attraction. An equally challenging criterion is to provide these substances in sufficient quantities (a few grams) to realize constant beams for a sufficient period of time to enable both alignment and interference experiments. While the requirement of 'large scale availability' constrains the variety of suitable structures, clever design and synthesis enabled us to push the mass limit to beyond 10 kDa [59]. Pulsed laser desorption of functionalized molecules from thin surfaces [87] has been shown to be more economical and applicable to an even larger variety of potential structures.

To minimize intermolecular attraction and to improve the volatility of molecules, their peripheral decoration with highly fluorinated alkyl chains is a successful strategy. The strong electron-withdrawing character of the fluorine atoms localizes the electron densities and decreases the electron mobility. The decoration with perfluorinated alkyl chains thus increases the mass of the target structure, while keeping the polarizability and induced dipole interaction low, a particularly appealing feature when 'heavy' particles are of interest. Furthermore, the stability of a C-F bond compares favorably with a C-H bond and the mass spectrum is kept clean, since fluorine is a monoisotopic element. The strategy has been successfully applied to a variety of model compounds ranging from simple dyes like azobenzenes [88], porphyrins [58, 89, 90], and phthalocyanines [45], to interlinked benzene subunits [91], and advanced molecular libraries pushing the limits of diffraction experiments [59, 67].

The feasibility of thermal peptide beams was studied using derivatives of a tryptophan-containing tripeptide. While an unprotected alanine-tryptophan-alanine (Ala-Trp-Ala) showed only fragments in the VUV-post-ionization mass spectrum ($\lambda = 157\,\text{nm}$), the intact molecular ion could be observed after removing internal charges and hydrogen bond donors by acetylation and amidation of the termini and methylation of the peptidic amide protons [92, 93]. The introduction of fluoroalkyl chains at the N-terminus or both termini improved the relative intensity of the molecular ion substantially despite the increase in molecular mass. The best results with the least fragmentation were obtained when fluoroalkyl chains were introduced, and the

N-methylation was omitted. Considerably more massive peptidic constructs could be launched and VUV-ionized under femtosecond laser desorption even reaching beyond 20 kDa for a 50 amino acid Trp-Lys construct which was extensively decorated with fluoroalkyl chains [65].

The second crucial factor that must be considered in the molecular design is the detection method. The observation of neutral molecules is challenging at low beam densities, and fragmentation-free post-ionization becomes generally more challenging with higher molecular mass [94]. The detectability, however, can be improved substantially by molecular design. The presence of a suitable chromophore enables the observation of individual molecules by fluorescence, which allowed real-time single-molecule imaging in far-field diffraction experiments [45]. Large, electron-rich π-systems are also attractive for photo-induced post-ionization [87]. Oligopeptides with tryptophan units turned out to be particularly suited for photoionization mass spectrometry because of the high absorption cross section of the indole subunit, the only group in any natural amino acid that is susceptible to single-photon ionization at 157 nm [66, 93]. A high tryptophan density even permitted the VUV-ionization of a peptidic construct of more than 20 kDa, thereby exceeding the mass limit for VUV-ionization by one order of magnitude over the previous standard for biomolecules [65]. Detection by mass spectrometry is appealing, as it eases the requirement of monodispersivity and thereby allows a new approach based on molecular libraries. These ensembles of molecules have different numbers of identical subunits, have a broader mass range, but with well-defined and well-separated masses.

3.2.2 Specific Examples

Fullerenes were used in the first diffraction experiments with organic molecules [42] and in many calibration experiments ever since. Their high thermal stability facilitates the creation of intense thermal beams from simple Knudsen cells, and they can be detected by electron impact ionization or by thermal ionization in an intense laser field [43, 95] followed by ion counting. Pure fullerene powder is also readily available in bulk quantities. Vapor pressures of about 0.1 hPa can be reached by heating the powder to 900 K, which generates an intense molecular beam with velocities in the range of 100–200 m/s.

Vitamins and provitamins such as α-tocopherol (vitamin E), β-carotene (provitamin A), 7-dehydrocholesterol (converted to provitamin D3 upon absorption of UV light) and phylloquinone (vitamin K1) have been interfered and deflected in the KDTLI experiment. Thermal sublimation of such fragile biomolecules always competes with fragmentation. At 500 K, the beam would typically last for only about 30 minutes.

Natural peptides do not evaporate or sublimate intact in a continuous thermal source. However, they can be launched by nanosecond [96] or femtosecond pulsed laser desorption sources, if they are immediately entrained into an adiabatically expanding seed gas. This recently enabled interference of a polypeptide with 15 amino acids,

Gramicidin A, in the OTIMA experiment. The fragility of large peptides, limited photoionization cross sections (needed for optical gratings and post-ionization mass spectrometry) of complex peptides, and carrier gas velocity, are the reasons why Gramicidin A is the most complex natural peptide in matter-wave experiments to date [66] (Fig.3a).

Thermal beams of functionalized tripeptides: Peptides are very fragile compounds—even simple dipeptides hardly survive the temperature needed to build up the vapor pressure required for molecular beam experiments. Interestingly, perfluoroalkly functionalization can facilitate the formation of thermal beams of even tripeptides [97] to a degree that matter wave interference became possible [93] (see Fig.6e).

Laser desorbed beams of large tailored polypeptides: The combination of fluoroalkyl decoration and ultrafast (femtosecond) laser desorption into a cold seed gas enabled launching even complex neutral peptides composed of up to 50 amino acids [65]. Their successful intact detection using single-photon ionization at 157 nm required optimizing the peptides for a tryptophan content as high as 50%.

Electrosprays of modified biopolymers: We have recently started investigating a novel approach to generating neutral biomolecular beams using bioconjugation techniques. Peptides with a photocleavable tag, introduced by amidation of surface amino groups with N-hydroxysuccinimide esters (NHS-esters) can be readily volatilized and ionized in an electrospray source. The emerging ions can be guided and manipulated using electric fields and they can even be cooled in a buffer gas. Subsequent neutralization can be achieved by selective photocleavage of the tag in an intense pulsed laser field. This mechanism has been demonstrated for various peptides [98] and even for human insulin [68].

Molecular libraries have been developed and optimized for high-mass interferometry. The concept is displayed in Fig.6f. A readily ionizable porphyrin architecture exposing numerous pentafluorophenyl groups was synthesized as a pure compound. In a subsequent aromatic nucleophilic substitution reaction fluorine atoms were substituted by highly fluorinated alkyl thiol chains. Since each reaction replaces exactly one fluorine atom by one fluorinated alkyl thiol chain, an entire molecular library emerges with precisely known masses, differing by the value of (M(fluorinated alkyl thiol chain)-M(FH)). Electron impact ionization mass spectrometry (EI-QMS) allowed the selective detection of a particular mass range.

Near-field interferometry tolerates a mass distribution even in excess of $\Delta m/m \simeq 10\,\%$ and neither isomers nor isotopes impair the experiment. Each molecule constitutes its own de Broglie wave and as long as the electromagnetic properties are similar the interference fringes will appear at the same position. Such a library was first built around a single tetrakispentafluorophenylporphyrin with 20 substitutable fluorine atoms [89] (see Fig. 6f), and successfully used in KDTL interferometry [59]. A dendritic porphyrin architecture with 60 substitutable fluorine atoms gave access to an even larger library that allowed us pushing the mass record in LUMI to beyond 25 kDa [67] (see Fig. 3b).

3.3 Molecule Interference Experiments

Numerous molecular properties have already been probed in quantum-assisted measurements. Here we restrict ourselves to experiments performed at the University of Vienna. They all rely on measurements of interference contrast and fringe deflection in generalized Talbot-Lau interferometers. Experimental results are divided into four categories; electronic, magnetic, and optical properties, as well as the measurement of inertial forces, as summarized in Table 1.

3.3.1 Electronic Properties

Electric deflection experiments require an electrode that provides a uniform force field. The transverse force on a polarizable particle (without permanent dipole moment) is then given by

$$F_x = -\nabla U_{ind} = -\nabla \frac{1}{2} \vec{d}_{ind} \cdot \vec{E} = \alpha_0 (E \cdot \nabla) E_x. \tag{3}$$

Here we include the possibility of a thermally induced dipole moment $d_{ind} = \alpha_0 E$. The factor of $1/2$ is due to the work done by the field inducing the dipole moment, and x is the direction transverse to both the molecular beam and to the grating bars.

Equation (3) shows that a field satisfying $(\vec{E} \cdot \nabla) E_x = \text{const}$ gives a constant transverse force proportional to the particles' static polarizability α_0. Electrodes have been designed and built with a tailored geometry to provide such a force, as described in Ref. [99] (see Fig. 4a).

Static polarizability: The first polarizability measurements in Talbot-Lau interferometry were made with fullerenes [101]. These measurements were repeated with improved precison and accuracy in the LUMI experiment [102]. Here, we took advantage of the ability to calibrate the setup in-situ with atomic cesium, the polarizability of which has been precisely measured with Mach-Zehnder interferomtery [35, 115] (see Fig. 5). Improving the precision even further, to better much than 1%, is only sensible for cold molecules, with improved control over the internal state.

Structural isomers have identical chemical composition and mass but different geometries and electronic properties (see Fig.6c). We consider two specially synthesized isomers which differ in their susceptibility by more than 20% because the molecule on the left of Fig.6c has a widely delocalized electron system, while electron delocalization is constrained to the phenyl rings in the molecule on the right. This can be easily distinguished in KDTLI deflectometry [91].

Dynamic dipole moment: While fullerenes are rigid, isotropic bodies, well-characterized by the scalar static polarizability α_0, this is not the case for floppy molecules such as the perfluoroalkyl-functionalized diazobenzenes [88] (see Fig. 6a). At a source temperature of 500 K these molecules undergo rapid conformational

Table 1 A list of quantum-assisted deflection experiments conducted in Vienna, sorted according to the type of measurement (electric, magnetic, optical, and inertial). Given are the particle properties measured, the interferometer in which the experiment was done, and the particle type with which it was done

Property	Interferometer	Particle
Electronic properties		
Static polarizability (*including dynamical electric dipole moment)	TLI, LUMI	C_{60}, C_{70} [101, 102]
	KDTLI	Perfluoroalkylated azobenzenes* [88]
	KDTLI	β-carotene*, vitamins E and K* [103]
	LUMI	Perfluoroalkylated tripeptides* [93]
Permanent dipole moment	KDTLI	Fe-TPP-Cl [104]
Fragment analysis	KDTLI	Perfluoroalkylated palladium complex and fragments [105]
Constitutional isomers	KDTLI	$C_{49}H_{16}F_{52}$ [91]
Magnetic properties		
Diamagnetic susceptibility, atoms	LUMI	Barium and strontium [82]
Diamagnetic susceptibility, molecules	LUMI	Anthracene and adamantane [106]
Optical properties		
Optical polarizability	KDTLI	C_{60}, C_{70} [107]
	KDTLI	C_{60}, C_{70}, $C_{60}F_{36}$, $C_{60}F_{48}$ [108]
	KDTLI	β-carotene, vitamins E and K [103]
Absorption cross section	KDTLI, grating interaction	C_{60}, C_{70}, $C_{60}F_{36}$, $C_{60}F_{48}$ [108, 109]
	KDTLI, recoil spectroscopy	C_{60} [110]
Conformer selection, proposed	Far-field, diffraction at dipole phase grating	Phenylethylamine [111]
Spectroscopy techniques, proposed	OTIMA, via recoil and depletion	[112]
Inertial forces		
Gravity/Weak equiv. princ.	OTIMA	TPP and derivatives [113]
Gravity-Coriolis compens.	LUMI	C_{60} [114]

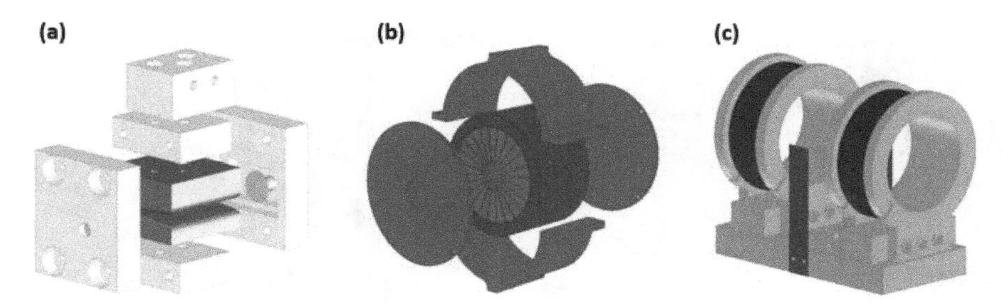

Fig. 4 Deflectors in our TLI, KDTLI and LUMI experiments. **a** A uniform $(\vec{E} \cdot \nabla)E_x$ field to measure polarizabilities and induced dipole moments [99]. **b** A modified Halbach array with a uniform $(\vec{B} \cdot \nabla)B_x$ to measure magnetic susceptibilities in LUMI [100]. **c** Anti-Helmholtz coils with a uniform ∇B_x for probing permanent magnetic moments in LUMI (image: S. Pedalino)

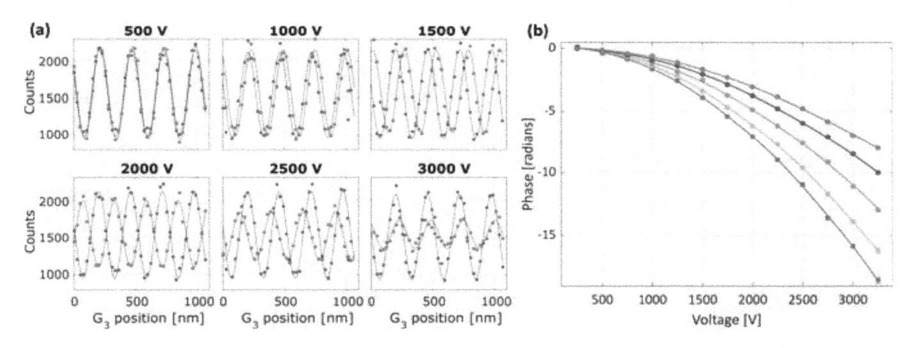

Fig. 5 **a** Electric deflection of C_{60} in the LUMI experiment. **b** The fringe shifts show the expected quadratic dependence on the electrode voltage for a selection of different molecular beam velocities

changes on the picosecond time scale, which leave the static and optical polarizability nearly constant, but may change the instantaneous electric dipole moment d_e by as much as 300%. This contributes to the net electronic susceptibility according to

$$\chi_{elec} = \alpha_0 + \frac{< d_e^2 >_T}{3k_B T}, \tag{4}$$

where the second term is due to the thermally averaged value of the dipole moment [116]. In the electric deflectometry experiments described here the total susceptibility is measured, and with the aid of ab initio molecular dynamics simulations the relative contributions of the static polarizability and the averaged thermally induced dipole moments can be extracted [88, 103]. It is interesting that de Broglie interferometry, which is primarily concerned with center-of-mass motion, can still reveal the influence of fast conformational changes through their influence on the molecules' response in an electric field. It is expected that molecular sequence isomers will exhibit different dynamic dipole moments and be separable in experiments with good signal-to-noise ratio [117].

(a) Time evolution of floppy azobenzenes

C_7F_{15} C_7F_{15}

t = 10 ns
d = 3.57 Debye

C_7F_{15} C_7F_{15}

t = 35 ns
d = 0.85 Debye

(b) Thermal fragmentation of a Pd catalyst precursor

C_8F_{17} Cl Pd P Cl

(c) Structure isomers: same composition, different geometry

C_6F_{13}

$\chi = 127 \times 4\pi\varepsilon_0 Å^3$

$\chi = 102 \times 4\pi\varepsilon_0 Å^3$

(d) Porphyrin derivatives with and without electric dipole moment

C_6H_5 N Cl Fe^{III}

d = 2.7 Debye

C_6H_5 N Fe^{II}

d = 0 Debye

(e) Functionalized tripeptides

C_5F_{11} NH C_8F_{17}

C_8F_{17} NH C_8F_{17}

(f) Molecular library based on porphyrin derivatives with varying numbers of substituents

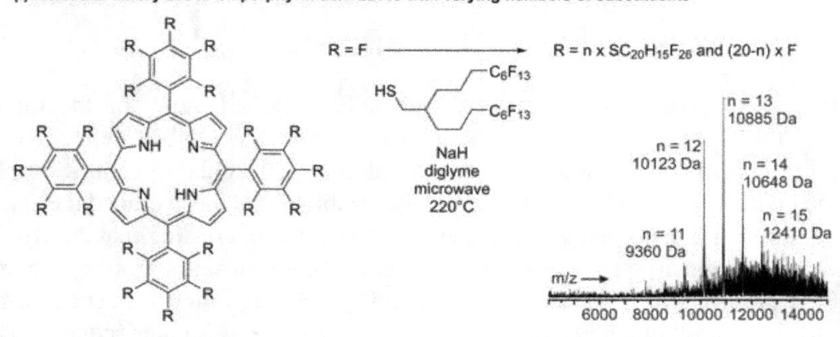

◀**Fig. 6** Interference-assisted deflectometry was used to elucidate the **dynamic dipole moments** of functionalized azobenzenes [88], **b** the **fragmentation** of a palladium catalyst precursor [105], **c electron delocalization** in constitutional isomers [91], and **d** permanent electric dipole moments in porpyhrin derivatives [104]. **e** Tripeptides optimized for both molecular beam formation and detection [93]. **f** Molecular libraries of members with well-defined molecular weight by random substitution of fluorine atoms with highly fluorinated alkyl chains [89]

Permanent dipole moment: Molecules with a permanent electric dipole moment experience a shift of their fringe pattern that depends upon the orientation of the molecule. Since most beam sources emit molecules with random initial orientation and in a mixture of thermally excited rotational states, each molecule experiences a different shift according to its orientation as it tumbles through the electric field. Molecules with a permanent electric dipole moment thus exhibit a reduced interference contrast, which can be used to distinguish them from polarizable particles with no permanent moment. This has been demonstrated with the porphyrin derivatives Fe-TPP and Fe-TPP-Cl (see Fig. 6d), which differ only by a single chlorine atom and a dipole moment of 2.7 D. Measuring the fringe deflection as a function of electrode voltage showed that both compounds have similar polarizabilities, while measuring the interference contrast revealed a much faster decay in contrast for the polar compound [104].

3.3.2 Magnetic Properties

Magnetic deflection, even more so than electric deflection, represents the huge impact of Otto Stern on the landscape of experimental physics, and has triggered a number of Stern-Gerlach type beam experiments [118–120]. However, only recently have similar experiments been performed in molecule interferometry.

The conceptual design of quantum-assisted magnetic deflectometry is identical to that of electric deflection. Here we aim to measure the magnetic susceptibility of particles subject to a uniform force which is introduced via a specially designed Halbach array of permanent magnets [100], as illustrated in Fig. 4b. The magnet can be translated in and out of the molecular beam, allowing us to take differential phase measurements referenced to a no-field situation.

In analogy to electric deflection, we require a region with

$$(\vec{B} \cdot \nabla) B_x = \text{const.} \tag{5}$$

such that magnetically susceptible particles with no permanent magnetic moment experience a uniform transverse force. Species with permanent magnetic moments, i.e. paramagnetic particles, will exhibit a reduced interference contrast, which is why this technique is best suited for measuring diamagnetic deflections or second order paramagnetic contributions (temperature-independent paramagnetism [116]).

The first measurements with the magnetic deflector described in Ref. [100] were performed in the LUMI experiment, taking advantage of the long interferometer baseline to observe the small diamagnetic deflection of thermal beams of the alkaline earth atoms barium and strontium [82]. The measured susceptibilities agreed well with the calculated values, and represent the first direct measurement of the ground-state diamagnetism of isolated particles. The sensitivity of the method was further illustrated by demonstrating the complete loss of interference contrast for the odd isotopes of barium and strontium which contain an unpaired nuclear spin, showing that even permanent moments on the order of a nuclear magneton are sufficient to completely dephase the interference fringes.

This work was recently extended to molecules [106], for which the situation is more complex than for atoms due to coupling with rotational states as well as alignment effects in the molecular beam. In this work, we measured the diamagnetic deflection of anthracene, a planar aromatic molecule, and adamantane, a tetrahedrally symmetric molecule. We observed the predicted isotropically averaged susceptibility for adamantane but a surprisingly large value for anthracene, which would be consistent with edge-on alignment of the planar molecules in the supersonic expansion. This alignment leads to the broadside orientation of anthracene being over-represented during its transit through the deflection region. Due to anthracene's aromaticity, the molecular plane has a significantly larger susceptibility tensor component than the other orientations, leading to the larger-than-isotropic observed deflection.

The difference to electric deflection experiments is apparent in the presence of a non-zero magnetic moment, as when there are unpaired nuclear or electron spins. In this case, as in the seminal Stern-Gerlach experiment, the quantization of the magnetic moment plays a role in the behavior of the particles in the magnetic field. When exposed to a $(\vec{B} \cdot \nabla)B_x$ field, the various projections will be deflected in different directions, leading to a reduction in interference contrast unless a spin state is selected and maintained in the interferometer (see Fig. 7). However, a constant ∇B_x field (such that the force on a permanent moment is uniform across the beam)

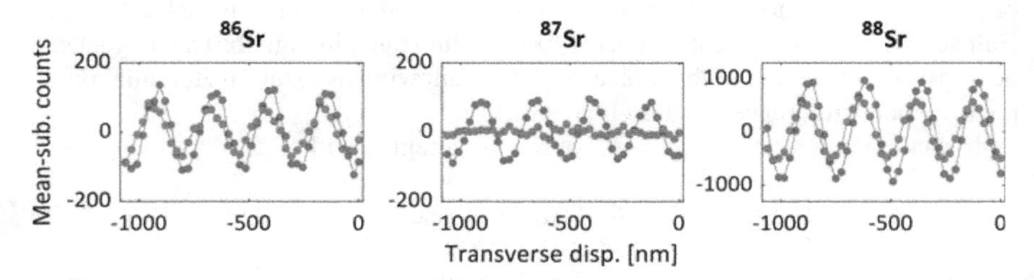

Fig. 7 In the LUMI experiment, we conducted the first beam deflection experiments to measure diamagnetic susceptibilities of atoms and molecules. In the first demonstration [82], the alkaline earth atoms barium and strontium were used in a TLI scheme. Here one can see both the diamagnetic deflection of the even isotopes of strontium (^{86}Sr and ^{88}Sr) and the complete washing out of the interference fringes of ^{87}Sr, which contains a small permanent magnetic moment due to an unpaired nuclear spin. Reference data is shown in blue and deflection data in red

can lead to revivals in interference visibility when the magnetic sub-levels are shifted by integer multiples of the grating period. This has been demonstrated in the LUMI experiment using anti-Helmholtz coils (illustrated in Fig. 4c) to observe the effect in atomic cesium [121], and experiments to observe the effect in triplet-excited fullerenes are in progress.

3.3.3 Optical Properties

There are several optical properties of molecules which can be extracted using quantum-assisted measurements. This can be accomplished by introducing an additional laser to the interferometer in analogy to the previously described deflectometry experiments and performing recoil spectroscopy, as proposed in Ref. [122]. The extraction of absolute absorption cross sections of dilute beams of C_{70} fullerenes was demonstrated using this technique in Ref. [110] in the KDTLI experiment.

Another approach to probe optical properties is to take advantage of the matter-light interactions which always occur in interferometer schemes with optical gratings. Since the contrast obtained in a KDTLI experiment depends on the AC polarizability of the molecule at the grating wavelength, this can be used as a measurement for optical polarizability, as done in Refs. [107, 108].

The sensitivity of the KDTLI scheme to optical polarizability can also be used to study molecular fragmentation [105]. The optical polarizability of a fragment of the palladium catalyst $C_{96}H_{48}C_{12}F_{102}P_2Pd$ was extracted by measuring the interference contrast as a function of the optical grating power. By comparing the measured polarizability to that of the intact particle versus the fragment, it could be determined whether the molecule fragmented already in the source, or only during the detection, after traversing the interferometer. Classical beam deflectometry could not have distinguished the origin of fragmentation, since the deflection depends only on the polarizability-to-mass ratio, which is nearly the same for the parent molecule and its fragments. Here, since the phase imprinted by the second grating depends only on the optical polarizability, it could be determined that the molecule fragmented in the source rather than the detector.

Measuring the optical polarizability is also useful for estimating the static polarizability of molecules, since for fullerenes and many other large organic molecules, we find that the static polarizability approximates the optical polarizability to within a few 10%. This is in agreement with the observation that most optical transitions for molecules in this complexity class are 30–50 nm wide.

There have been several proposals for other ways to utilize the sensitive dependence of molecule diffraction on optical interactions. Several spectroscopy setups have been proposed in the context of the OTIMA experiment, including multi-photon recoil and polarizability spectroscopy [112]. In the far-field diffraction experiment, a near-resonant ultraviolet optical grating could potentially be used for efficient sorting of conformers [111]. It has also been proposed to employ optical helicity fringes to create a diffraction grating that discriminates chiral enantiomers [123].

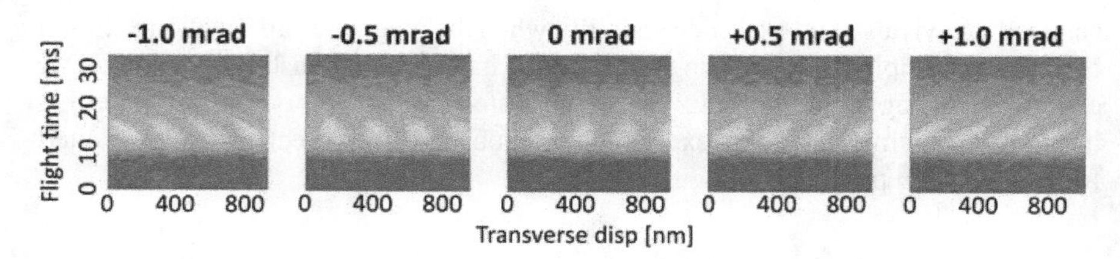

Fig. 8 Time-resolved interference scans showing the competing effects of gravitational and Coriolis phase shifts as a function of the interferometer roll angle. On the far left gravity is responsible for the large shearing, while on the far right the gratings are nearly aligned with gravity and the shearing is due to the Coriolis force. In Ref. [114] these two phase shifts were used to passively compensate one another, but a similar technique could be used for the purpose of measuring gravitational or Sagnac phases directly

3.3.4 Inertial Forces

Interferometers have long been used as inertial sensors, from Sagnac loop interferometers with light to sensitive gravity and gravity gradient sensors made with atom interferometers. Molecule interferometers do not compete in sensitivity due to the comparatively poor signal-to-noise ratios and smaller enclosed areas, but they are still sensitive to such effects. In Ref. [114], the competing phase shifts of fullerenes due to the Coriolis effect and gravity were mapped as a function of velocity for different roll angles of the interferometer setup (see Fig. 8). Molecule interferometry also enables weak equivalence principle measurements of a wider variety of species and internal energies and properties than atom interferometry experiments. This has been demonstrated in the OTIMA experiment by comparing the gravitational phase shift of various isotopomeres of tetraphenylporphyrin [113].

4 Outlook

Molecule interferometry and deflectometry have been inspired by work that was started by Otto Stern 100 years ago. Much of the research in the field since then can be seen as a very extended footnote to the ideas of Otto Stern. And yet we foresee years of exciting research in the attempt to push matter-wave interferometry to ever higher mass and complexity, and to further explore the interface to chemistry, biology and the classical world.

Otto Stern remarked in several of his writings on the particular challenge of preparing molecular beams. This is where quantum optics and chemistry have found a very fruitful overlap and where we still expect thrilling developments: the tailoring of molecules to the needs of quantum optics as well as the use of quantum optics to retrieve information about molecules is a new field of research that opens promising perspectives.

Acknowledgements We acknowledge funding by the European Research Council (Project No. 320694), the Austrian Science Funds (FWF P-30176, P-32543-N), the Swiss National Funds (Project No. 200020159730), the SNI PhD School (P1403) as well as the tireless contribution of many master and PhD students as well as postdocs who contributed over the years to various molecule interferometers and interferometer applications for molecular science.

References

1. O. Stern, Zeitschrift für Physik **39**, 751–763 (1926)
2. E. Reiger, L. Hackermüller, M. Berninger, M. Arndt, Opt. Commun. **264**(2), 326 (2006)
3. C. Davisson, L.H. Germer, Phys. Rev. **30**, 705 (1927)
4. I. Estermann, O. Stern, Z. Phys. **61**, 95 (1930)
5. P.E. Moskowitz, P.L. Gould, S.R. Atlas, D.E. Pritchard, Phys. Rev. Lett. **51**, 370 (1983)
6. P.L. Gould, G.A. Ruff, D.E. Pritchard, Phys. Rev. Lett. **56**, 827 (1986)
7. D.W. Keith, M.L. Schattenburg, H.I. Smith, D.E. Pritchard, Phys. Rev. Lett. **61**, 1580 (1988)
8. D.W. Keith, C.R. Ekstrom, Q.A. Turchette, D.E. Pritchard, Phys. Rev. Lett. **66**(21), 2693 (1991)
9. C.J. Bordé, Phys. Lett. A **140**, 10 (1989)
10. C.J. Bordé, S. Avrillier, A. Van Lerberghe, C. Salomon, D. Bassi, G. Scoles, J. Phys. Coll. **42**(C8), 15 (1981)
11. M. Kasevich, D.S. Weiss, E. Riis, K. Moler, S. Kasapi, S. Chu, Phys. Rev. Lett. **66**, 2297 (1991)
12. A. Peters, K. Yeow-Chung, S. Chu, Nature **400**, 849 (1999)
13. G. Lamporesi, A. Bertoldi, L. Cacciapuoti, M. Prevedelli, G. Tino, Phys. Rev. Lett. **100**, 5 (2008)
14. P. Asenbaum, C. Overstreet, M. Kim, J. Curti, M.A. Kasevich, arXiv:2005.11624v1 (2020)
15. R.H. Parker, C. Yu, W. Zhong, B. Estey, H. Müller, Science **360**(6385), 191 (2018)
16. I. Dutta, D. Savoie, B. Fang, B. Venon, C.L. Garrido Alzar, R. Geiger, A. Landragin, Phys. Rev. Lett. **116**(18), 183003 (2016)
17. D. Savoie, M. Altorio, B. Fang, L.A. Sidorenkov, R. Geiger, A. Landragin, Sci. Adv. **4**, 7948 (2018)
18. W. Chaibi, R. Geiger, B. Canuel, A. Bertoldi, A. Landragin, P. Bouyer, Phys. Rev. D **93**(2), 021101 (2016)
19. Y.A. El-Neaj, C. Alpigiani, S. Amairi-Pyka, H. Araújo, A. Balaž, A. Bassi, L. Bathe-Peters, B. Battelier, A. Belić, E. Bentine, J. Bernabeu, A. Bertoldi, R. Bingham, D. Blas, V. Bolpasi, K. Bongs, S. Bose, P. Bouyer, T. Bowcock, W. Bowden, O. Buchmueller, C. Burrage, X. Calmet, B. Canuel, L.I. Caramete, A. Carroll, G. Cella, V. Charmandaris, S. Chattopadhyay, X. Chen, M.L. Chiofalo, J. Coleman, J. Cotter, Y. Cui, A. Derevianko, A. De Roeck, G.S. Djordjevic, P. Dornan, M. Doser, I. Drougkakis, J. Dunningham, I. Dutan, S. Easo, G. Elertas, J. Ellis, M. El Sawy, F. Fassi, D. Felea, C.H. Feng, R. Flack, C. Foot, I. Fuentes, N. Gaaloul, A. Gauguet, R. Geiger, V. Gibson, G. Giudice, J. Goldwin, O. Grachov, P.W. Graham, D. Grasso, M. van der Grinten, M. Gündogan, M.G. Haehnelt, T. Harte, A. Hees, R. Hobson, J. Hogan, B. Holst, M. Holynski, M. Kasevich, B.J. Kavanagh, W. von Klitzing, T. Kovachy, B. Krikler, M. Krutzik, M. Lewicki, Y.H. Lien, M. Liu, G.G. Luciano, A. Magnon, M.A. Mahmoud, S. Malik, C. McCabe, J. Mitchell, J. Pahl, D. Pal, S. Pandey, D. Papazoglou, M. Paternostro, B. Penning, A. Peters, M. Prevedelli, V. Puthiya-Veettil, J. Quenby, E. Rasel, S. Ravenhall, J. Ringwood, A. Roura, D. Sabulsky et al., EPJ Quant. Technol. **7**, 1 (2020)
20. C. Burrage, E.J. Copeland, E.A. Hinds, J. Cosmol. Astroparticle Phys. **2015**(03), 042 (2015)
21. P. Hamilton, M. Jaffe, P. Haslinger, Q. Simmons, H. Müller, J.T. Khoury, Science **349**, 849 (2015)
22. P.R. Berman, B. Dubetsky, Phys. Rev. A **59**, 2269 (1999)
23. A.D. Cronin, J. Schmiedmayer, D.E. Pritchard, Rev. Mod. Phys. **81**(3), 1051 (2009)
24. G. Tino, M. Kasevich, Atom interferometry, in *Proceedings of the International School of Physics "Enrico Fermi"*, vol. 188 (IOS, Varenna, 2014)

25. M.H. Anderson, J.R. Ensher, M.R. Matthews, C.E. Wieman, E.A. Cornell, Science **269**, 198 (1995)
26. K.B. Davis, M.O. Mewes, M.R. Andrews, N.J. van Druten, D.S. Durfee, D.M. Kurn, W. Ketterle, Phys. Rev. Lett. **75**, 3969 (1995)
27. J. Herbig, T. Kraemer, M. Mark, T. Weber, C. Chin, H.C. Nagerl, R. Grimm, Science **301**, 1510 (2003)
28. C. Kohstall, S. Riedl, E.R. Sánchez Guajardo, L.A. Sidorenkov, J. Hecker Denschlag, R. Grimm, New. J. Phys. **13**(6), 065027 (2011)
29. W. Schöllkopf, J.P. Toennies, Science **266**, 1345 (1994)
30. M.S. Chapman, T.D. Hammond, A. Lenef, J. Schmiedmayer, R.A. Rubenstein, E. Smith, D.E. Pritchard, Phys. Rev. Lett. **75**, 3783 (1995)
31. C. Bordé, N. Courtier, F.D. Burck, A. Goncharov, M. Gorlicki, Phys. Lett. A **188**, 187 (1994)
32. B.S. Zhao, G. Meijer, W. Schoellkopf, Science **331**(6019), 892 (2011)
33. C. Ekstrom, J. Schmiedmayer, M. Chapman, T. Hammond, D. Pritchard, Phys. Rev. A **51**(5), 3883 (1995)
34. A. Miffre, M. Jacquey, M. Büchner, G. Trenec, J. Vigue, Phys. Rev. A **73**, 011603(R) (2006)
35. M.D. Gregoire, I. Hromada, W.F. Holmgren, R. Trubko, A.D. Cronin, Phys. Rev. A **92**, 5 (2015)
36. J. Schmiedmayer, M. Chapman, C. Ekstrom, T. Hammond, S. Wehinger, D. Pritchard, Phys. Rev. Lett. **74**(7), 1043 (1995)
37. V.P.A. Lonij, Atom optics, core electrons, and the van der Waals potential. Thesis (2011)
38. R. Trubko, J. Greenberg, M.T.S. Germaine, M.D. Gregoire, W.F. Holmgren, I. Hromada, A.D. Cronin, Phys. Rev. Lett. **114**, 14 (2015)
39. A. Fallon, C. Sackett, Atoms **4**, 2 (2016)
40. C. Lisdat, M. Frank, H. Knöckel, M.L. Almazor, E. Tiemann, Eur. Phys. J. D **12**, 235 (2000)
41. P. Rousseau, H. Khemliche, A.G. Borisov, P. Roncin, Phys. Rev. Lett. **98**, 1 (2007)
42. M. Arndt, O. Nairz, J. Voss-Andreae, C. Keller, G. van der Zouw, A. Zeilinger, Nature **401**, 680 (1999)
43. D. Ding, J. Huang, R. Compton, C. Klots, R. Haufler, Phys. Rev. Lett. **73**(8), 1084 (1994)
44. O. Nairz, M. Arndt, A. Zeilinger, Am. J. Phys. **71**(4), 319 (2003)
45. T. Juffmann, A. Milic, M. Müllneritsch, P. Asenbaum, A. Tsukernik, J. Tüxen, M. Mayor, O. Cheshnovsky, M. Arndt, Nature Nanotech. **7**, 297 (2012)
46. W.H.F. Talbot, Philos. Mag. **9**, 401 (1836)
47. E. Lau, Ann. Phys. **6**, 417 (1948)
48. J. Clauser, *De Broglie-Wave Interference of Small Rocks and Live Viruses* (Kluwer Academic, 1997), pp. 1–11
49. B. Brezger, L. Hackermüller, S. Uttenthaler, J. Petschinka, M. Arndt, A. Zeilinger, Phys. Rev. Lett. **88**, 100404 (2002)
50. L. Hackermüller, S. Uttenthaler, K. Hornberger, E. Reiger, B. Brezger, A. Zeilinger, M. Arndt, Phys. Rev. Lett. **91**(9), 090408 (2003)
51. T. Juffmann, S. Truppe, P. Geyer, A.G. Major, S. Deachapunya, H. Ulbricht, M. Arndt, Phys. Rev. Lett. **103**, 26 (2009)
52. S. Gerlich, L. Hackermüller, K. Hornberger, A. Stibor, H. Ulbricht, M. Gring, F. Goldfarb, T. Savas, M. Müri, M. Mayor, M. Arndt, Nat. Phys. **3**(10), 711 (2007)
53. C. Brand, M. Sclafani, C. Knobloch, Y. Lilach, T. Juffmann, J. Kotakoski, C. Mangler, A. Winter, A. Turchanin, J. Meyer, O. Cheshnovsky, M. Arndt, Nat. Nanotechnol. **10**, 845 (2015)
54. P.L. Kapitza, P.A.M. Dirac, Proc. Camb. Philos. Soc. **29**, 297 (1933)
55. D.L. Freimund, K. Aflatooni, H. Batelaan, Nature **413**, 142 (2001)
56. O. Nairz, B. Brezger, M. Arndt, A. Zeilinger, Phys. Rev. Lett. **87**, 160401 (2001)
57. B. Brezger, M. Arndt, A. Zeilinger, J. Opt. B. **5**, 82 (2003)
58. S. Gerlich, S. Eibenberger, M. Tomandl, S. Nimmrichter, K. Hornberger, P. Fagan, J. Tüxen, M. Mayor, M. Arndt, Nat. Commun. **2**, 263 (2011)
59. S. Eibenberger, S. Gerlich, M. Arndt, M. Mayor, J. Tüxen, Phys. Chem. Chem. Phys. **15**, 14696 (2013)
60. S. Nimmrichter, K. Hornberger, P. Haslinger, M. Arndt, Phys. Rev. A **83**, 043621 (2011)

61. P. Haslinger, N. Dörre, P. Geyer, J. Rodewald, S. Nimmrichter, M. Arndt, Nat. Phys. **9**, 144–148 (2013)
62. R. Abfalterer, S. Bernet, C. Keller, M. Oberthaler, J. Schmiedmayer, A. Zeilinger, Act. Phys. Slov. **47**(3/4), 165 (1997)
63. S. Fray, C.A. Diez, T.W. Hänsch, M. Weitz, Phys. Rev. Lett. **93**, 24 (2004)
64. N. Dörre, J. Rodewald, P. Geyer, B. von Issendorff, P. Haslinger, M. Arndt, Phys. Rev. Lett. **113**, 233001 (2014)
65. J. Schätti, P. Rieser, U. Sezer, G. Richter, P. Geyer, G.G. Rondina, D. Häussinger, M. Mayor, A. Shayeghi, V. Köhler, M. Arndt, Commun. Chem. **1**(1), 93 (2018)
66. A. Shayeghi, P. Rieser, G. Richter, U. Sezer, J. Rodewald, P. Geyer, T. Martinez, M. Arndt, Nat. Commun. **11**, 144 (2020)
67. Y.Y. Fein, P. Geyer, P. Zwick, F. Kiałka, S. Pedalino, M. Mayor, S. Gerlich, M. Arndt, Nat. Phys. (2019)
68. J. Schätti, M. Kriegleder, M. Debiossac, M. Kerschbaum, P. Geyer, M. Mayor, M. Arndt, V. Köhler, Chem. Commun. (Camb) **55**(83), 12507 (2019)
69. F. Kiałka, B. Stickler, K. Hornberger, Y.Y. Fein, P. Geyer, L. Mairhofer, S. Gerlich, M. Arndt, Physica Scripta (2018)
70. R. Schäfer, S. Schlecht, J. Woenckhaus, J.A. Becker, Phys. Rev. Lett. **76**(3), 471 (1996)
71. K. Bonin, V. Kresin, *Electric-Dipole Polarizabilities of Atoms, Molecules and Clusters* (World Scientific, 1997)
72. R. Antoine, I. Compagnon, D. Rayane, M. Broyer, P. Dugourd, N. Sommerer, M. Rossignol, D. Pippen, F.C. Hagemeister, M.F. Jarrold, Anal. Chem. **75**, 5512 (2003)
73. W.A. de Heer, V.V. Kresin, *Electric and Magnetic Dipole Moments of Free Nanoclusters* (CRC Press, 2011), pp. 10/1–13
74. T.M. Fuchs, R. Schäfer, Phys. Rev. A **98**, 6 (2018)
75. F. Filsinger, J. Kupper, G. Meijer, J.L. Hansen, J. Maurer, J.H. Nielsen, L. Holmegaard, H. Stapelfeldt, Angew Chem. Int. Ed. Engl. **48**(37), 6900 (2009)
76. Y.P. Chang, K. Dlugolecki, J. Küpper, D. Rösch, D. Wild, S. Willitsch, Science **342**(6154), 98 (2013)
77. E. Gershnabel, M. Shapiro, I. Averbukh, J. Chem. Phys. **135**(19), 194310 (2011)
78. M. Abd El Rahim, R. Antoine, L. Arnaud, M. Barbaire, M. Broyer, C. Clavier, I. Compagnon, P. Dugourd, J. Maurelli, D. Rayane, Rev. Sci. Instrum. **75**(12), 5221 (2004)
79. M.K. Oberthaler, S. Bernet, E.M. Rasel, J. Schmiedmayer, A. Zeilinger, Phys. Rev. A **54**, 3165 (1996)
80. J.F. Clauser, S. Li, Phys. Rev. A **49**, R2213 (1994)
81. K. Hornberger, S. Gerlich, P. Haslinger, S. Nimmrichter, M. Arndt, Rev. Mod. Phys. **84**, 157 (2012)
82. Y.Y. Fein, A. Shayeghi, L. Mairhofer, F. Kiałka, P. Rieser, P. Geyer, S. Gerlich, M. Arndt, Phys. Rev. X **10**, 011014 (2020)
83. M.A. Bouchiat, C. Bouchiat, Phys. Rev. A **83**, 5 (2011)
84. K. Zeiske, G. Zinner, F. Riehle, J. Helmcke, Appl. Phys. B **60**, 205 (1995)
85. E. Cohen, H. Larocque, F. Bouchard, F. Nejadsattari, Y. Gefen, E. Karimi, Nat. Rev. Phys. **1**(7), 437 (2019)
86. S. Lepoutre, A. Gauguet, M. Büchner, J. Vigué, Phys. Rev. A **88**, 4 (2013)
87. U. Sezer, L. Wörner, J. Horak, L. Felix, J. Tüxen, C. Götz, A. Vaziri, M. Mayor, M. Arndt, Anal. Chem. **87**, 5614–5619 (2015)
88. M. Gring, S. Gerlich, S. Eibenberger, S. Nimmrichter, T. Berrada, M. Arndt, H. Ulbricht, K. Hornberger, M. Müri, M. Mayor, M. Böckmann, N.L. Doltsinis, Phys. Rev. A **81**, 031604(R) (2010)
89. J. Tüxen, S. Eibenberger, S. Gerlich, M. Arndt, M. Mayor, Eur. J. Organ. Chem. (25), 4823 (2011)
90. P. Schmid, F. Stöhr, M. Arndt, J. Tüxen, M. Mayor, J. Am. Soc. Mass Spectrom. **24**(4), 602 (2013)
91. J. Tüxen, S. Gerlich, S. Eibenberger, M. Arndt, M. Mayor, Chem. Commun. **46**(23), 4145 (2010)

92. B.C. Das, S.D. Gero, E. Lederer, Biochem. Biophys. Res. Commun. **29**(2), 211 (1967)
93. J. Schätti, V. Köhler, M. Mayor, Y.Y. Fein, P. Geyer, L. Mairhofer, S. Gerlich, M. Arndt, J. Mass Spectrom. **55**(6), e4514 (2020). E4514 JMS-19-0196.R2
94. A. Akhmetov, J.F. Moore, G.L. Gasper, P.J. Koin, L. Hanley, J. Mass Spectrom. **45**(2), 137 (2010)
95. O. Nairz, M. Arndt, A. Zeilinger, J. Modern Opt. **47**(14–15), 2811 (2000)
96. M. Marksteiner, P. Haslinger, M. Sclafani, H. Ulbricht, M. Arndt, J. Phys. Chem. A **113**(37), 9952 (2009)
97. J. Schätti, U. Sezer, S. Pedalino, J.P. Cotter, M. Arndt, M. Mayor, V. Köhler, J. Mass Spectrom. **52**, 550 (2017)
98. M. Debiossac, J. Schätti, M. Kriegleder, P. Geyer, A. Shayeghi, M. Mayor, M. Arndt, V. Köhler, Phys. Chem. Chem. Phys. **20**, 11412 (2018)
99. A. Stefanov, M. Berninger, M. Arndt, Meas. Sci. Technol. **19**, 5 (2008)
100. L. Mairhofer, S. Eibenberger, A. Shayeghi, M. Arndt, Entropy **20**, 516 (2018)
101. M. Berninger, A. Stefanov, S. Deachapunya, M. Arndt, Phys. Rev. A **76**, 013607 (2007)
102. Y.Y. Fein, P. Geyer, F. Kiałka, S. Gerlich, M. Arndt, Phys. Rev. Res. **1**, 033158 (2019)
103. L. Mairhofer, S. Eibenberger, J.P. Cotter, M. Romirer, A. Shayeghi, M. Arndt, Angew. Chem. Int. Ed. **56**, 10947 (2017)
104. S. Eibenberger, S. Gerlich, M. Arndt, J. Tüxen, M. Mayor, New J. Phys. **13**(4), 043033 (2011)
105. S. Gerlich, M. Gring, H. Ulbricht, K. Hornberger, J. Tüxen, M. Mayor, M. Arndt, Angew Chem. Int. Ed. Engl. **47**(33), 6195 (2008)
106. Y.Y. Fein, A. Shayeghi, F. Kiałka, P. Geyer, S. Gerlich, M. Arndt, Phys. Chem. Chem. Phys. pp. 14,036–14,041 (2020)
107. L. Hackermüller, K. Hornberger, S. Gerlich, M. Gring, H. Ulbricht, M. Arndt, Appl. Phys. B **89**(4), 469 (2007)
108. K. Hornberger, S. Gerlich, H. Ulbricht, L. Hackermüller, S. Nimmrichter, I. Goldt, O. Boltalina, M. Arndt, New J. Phys. **11**, 043032 (2009)
109. J.P. Cotter, S. Eibenberger, L. Mairhofer, X. Cheng, P. Asenbaum, M. Arndt, K. Walter, S. Nimmrichter, K. Hornberger, Nat. Commun. **6**, 7336 (2015)
110. S. Eibenberger, X. Cheng, J.P. Cotter, M. Arndt, Phys. Rev. Lett. **112**, 250402 (2014)
111. C. Brand, B.A. Stickler, C. Knobloch, A. Shayegh, K. Hornberger, M. Arndt, Phys. Rev. Lett. **121**, 173002 (2018)
112. J. Rodewald, P. Haslinger, N. Dörre, B.A. Stickler, A. Shayeghi, K. Hornberger, M. Arndt, Appl. Phys. B **123**(1), 3 (2017)
113. J. Rodewald, N. Dörre, A. Grimaldi, P. Geyer, L. Felix, M. Mayor, A. Shayeghi, M. Arndt, New J. Phys. **20**, 033016 (2018)
114. Y.Y. Fein, F. Kiałka, P. Geyer, S. Gerlich, M. Arndt, New J. Phys. **22**, 033013 (2020)
115. M. Gregoire, N. Brooks, R. Trubko, A. Cronin, Atoms **4**(3), 21 (2016)
116. J.V. Vleck, *The Theory of Electric and Magnetic Susceptibilities* (Oxford University Press London, 1965)
117. H. Ulbricht, M. Berninger, S. Deachapunya, A. Stefanov, M. Arndt, Nanotechnology **19**, 045502 (2008)
118. W.D. Knight, R. Monot, E.R. Dietz, A.R. George, Phys. Rev. Lett. **40**, 1324 (1978)
119. U. Rohrmann, R. Schafer, Phys. Rev. Lett. **111**(13), 133401 (2013)
120. O. Amit, Y. Margalit, O. Dobkowski, Z. Zhou, Y. Japha, M. Zimmermann, M.A. Efremov, F.A. Narducci, E.M. Rasel, W.P. Schleich, R. Folman, Phys. Rev. Lett. **123**, 083601 (2019)
121. Y.Y. Fein, Long-baseline universal matter-wave interferometry. Thesis (2020)
122. S. Nimmrichter, K. Hornberger, Phys. Rev. A **78**, 023612 (2008)
123. R.P. Cameron, S.M. Barnett, A.M. Yao, New J. Phys. **16** (2014)

6

Quantum or Classical Perception of Atomic Motion

John S. Briggs

Abstract An assessment is given as to the extent to which pure unitary evolution, as distinct from environmental decohering interaction, can provide the transition necessary for an observer to perceive quantum dynamics as classical. This has implications for the interpretation of quantum wavefunctions as a characteristic of ensembles or of single particles and the related question of wavefunction "collapse". A brief historical overview is presented as well as recent emphasis on the role of the semi-classical "imaging theorem" in describing quantum to classical unitary evolution.

1 Introduction

In describing the motion of the silver atom beam through their apparatus in 1922, Gerlach and Stern [1] naturally assumed classical mechanics; only the internal angular momentum was quantised. Later, following Schrödinger's 1926 introduction of wave mechanics, more attention was given to the quantum state of motion of the atoms. For example, Bohm in 1951 [2] described the beam translational motion by quantum mechanics and gave particular attention to the question of interference between the waves of the two beams leaving the magnet region. An extensive discussion of the classical and quantum aspects is given in Gottfried and Yan [3] in terms of particle wave packets.

Here, the general description of the motion of atomic particles over macroscopic distances is considered. It is shown how classical features appear autonomously and that the perception of classical or quantal behaviour depends upon the extent and accuracy of detection of such motion. This has consequences for the meaning assigned to a wave function and to its interpretation as to describing the motion of a single particle or the motions of an ensemble of identical particles.

J. S. Briggs (✉)
Institute of Physics, University of Freiburg, Freiburg, Germany
e-mail: briggs@physik.uni-freiburg.de

Department of Physics, Royal University of Phnom Penh, Phnom Penh, Cambodia

1.1 Particle or Wave or Particle Ensemble?

In the scattering of electromagnetic waves around a sharp material object, the nature of the perceived outline depends upon the resolution and sensitivity of the instrument. For visible light, in the case of the human eye, usually a sharp outline of the object ascribable to a ray description of the light would be inferred. However, from measurement with an instrument able to resolve at sub-wavelength accuracy, a blurred outline corresponding to a diffraction pattern and ascribable to the wave nature of the light would be inferred. The instrument resolution is understood as the accuracy of position location. Important also is the sensitivity of the instrument, whose limit is taken here as the ability or not to register reception of a single quantum particle e.g. a photon, electron, atom or molecule.

If the detector sensitivity is sufficient one can monitor the arrival of individual particles. In the case of photons, their arrival at a screen appears in a seemingly arbitrary pattern until enough photons are counted. Then the statistical distribution gradually assumes the structured diffraction or interference form expected on the basis of the classical wave picture of electromagnetism. This is the wave-particle duality of light. With increasing resolution and sensitivity of the measurement, there are three levels of perception, classical ray trajectory, the wave picture and the ensemble of (quantum) particles picture.

A similar situation arises in the wave-particle duality of matter. For an ensemble of identical particles with a mass which is very large on an atomic scale, one assigns to their motion a unique classical trajectory so long as the resolution of, say position detection, is not itself on the atomic scale. When the mass of the particles *is* on the atomic scale it is necessary to calculate the average of their motion from the wave picture of quantum mechanics. Increasing the sensitivity to detect individual particles leads to a seemingly arbitrary pattern until the statistics are sufficient that a pattern predicted by the wave description emerges, for an experiment with electrons see Ref. [4]. Again there are three levels of perception of the ensemble; unique classical trajectory, many particles registered as a wave pattern or the statistical pattern from individual quantum particles. Which description is appropriate depends both upon the the resolution and the sensitivity of the measurement.

Indeed the analogy between the classical wave equations of electromagnetism (Helmholtz equations and paraxial approximation) and the wave equations of quantum mechanics (time-independent and time-dependent Schrödinger equations) is very close mathematically. This leads to the similarity of perception alluded to above. The semi-classical limit of quantum mechanics, used extensively below, corresponds to the eikonal approximation for electric wave propagation. Thus the quantum to classical limit for material particles corresponds to the wave to beam limit of electric field propagation. In optics, the large separation between source and observer is used to derive the Fraunhofer diffraction formula which is in close analogy to the asymptotic "Imaging Theorem" of quantum mechanics derived in section III.

One must be clear on this point. The quantum to classical transition in particle dynamics is the transition from wave to particle perception. Paradoxically, the quantum to classical transition in photon dynamics is the opposite, from particle to wave perception. It is the *wave to beam* transition of classical optical dynamics which is the analogue of the *quantum to classical* transition in particle dynamics.

A key element of quantum mechanics, not present for classical light, is the interpretation of the modulus squared of the wavefunction as a statistical probability. Here, two points of view have emerged. The first, to be called the ensemble picture, is that the probability describes the percentage of members of an ensemble of identical, and identically-prepared, particles having a particular value of a dynamical variable. The second, to be called the single-particle (SP) picture, considers that it is the probability with which an individual particle exhibits a given value out of the totality of possibilities. That is, on measurement the wavefunction "collapses" into one eigenstate of the observable with one definite value of the variable observed. The difference is that in the ensemble interpretation, only measurements on the whole ensemble are meaningful. In the SP picture meaning is assigned to a measurement on a single particle.

Here it is argued that only the ensemble interpretation of the wavefunction is tenable. However, it should be made clear from the outset that "ensemble" refers to an ensemble of N *measurements*, not necessarily N particles. The initial conditions have to be identical and the wavefunction gives statistical information on the outcomes. The measurements can be simultaneous or sequential. In the case of N particles the particles must be indistinguishable. This specification of ensembles of measurements is necessary since, unthinkable to the founders of quantum mechanics, experiments today can be made on trapped single electrons, atoms or molecules. Then the wavefunction gives only statistical information on a sequence of measurements in which the same particle initially is brought into the same state e.g. experiments on quantum jumps.

Furthermore, if the wavefunction refers to an entangled state of several particles, then that group of particles is to be regarded as a single member (single "particle") of the ensemble. Correspondingly, a feature that is very important but has been often neglected in the past, is the occurrence or otherwise of many-particle *good quantum numbers* describing eigenstates of some many-particle mechanical variable. This is because, for this special case, the measurement of the corresponding many-particle mechanical variable gives the same sharp value for all members of the ensemble. For the simple case of single-particle ensembles in an eigenstate, there is no difference between the SP and ensemble pictures, although a repeated sequence of measurements is still required to confirm the eigenstate unless defined uniquely by the preparation.

The object of this work is to re-appraise, in the light of the SP and ensemble pictures, the transition from quantum to classical mechanics by emphasising the role of the "Imaging Theorem" (IT) [5–10] in determining what an observer perceives as a consequence of the experimental resolution, sensitivity and information extraction. The IT was proved as long ago as 1937 by Kemble [5] whose aim was to show how a particle linear momentum vector could be measured and assigned in a collision experiment. Although largely forgotten until recently, here a more fundamental consequence of this theorem with regard to the quantum to classical transition is suggested, following Ref. [6].

The IT shows that any system of particles emanating from microscopic separations describable by quantum dynamics will acquire characteristics of classical trajectories simply through unitary propagation to the macroscopic separations at which measurements are made. Specifically, the IT equates the final position wavefunction $\Psi(\boldsymbol{r}_f, t_f)$ at a detector at *macroscopic* position and time, to the initial momentum wavefunction $\tilde{\Psi}(\boldsymbol{p}_i, t_i)$ at *microscopic* position and time but, importantly, where the variables $\boldsymbol{r}, \boldsymbol{p}$ and t are related by a *classical trajectory*. This justifies the standard approach of experimentalists who use classical trajectories to trace the motion of particles from reaction zone to detector, even though the particle correlations indicate existence of a quantum wavefunction. In fact the relevance of great advances in multi-particle coincident detection [11] to the criteria for quantum or classical perception given here cannot be underestimated.

The IT connects the *momentum* wavefunction in the microscopic collision zone with the *position* wavefunction at macroscopic distance. The initial position in the collision zone cannot be defined precisely but, for free motion, the initial momentum is conserved out to the detector. This feature of the IT is completely in correspondence with the supposition of Schmidt-Böcking et.al. in Chap. 12 of this volume [12], who demonstrate that the momentum of particles emanating from a microscopic collision can be determined with sub-atomic precision but the initial position can never be measured with comparable precision.

As emphasised in the chapter by Schmidt-Böcking et al. the ensemble picture is essential for the correct interpretation of the uncertainty relation (UR) for position and momentum. According to the Robertson formulation of the UR [13] the state-dependent spread in the measurement of position and momentum refer to ensemble averages. Therefore, by definition, the individual measurements must have an *accuracy* much less than the spread. The fact that, in an individual single measurement, the product of accuracies of simultaneous momentum and position measurements can be less than $\hbar$ is demonstrated convincingly in Chap. 12.

The IT defines the asymptotic wavefunction of any quantum complex after interacting with particle or photon beams in a microscopic collision region. Such collisions *always* result in entangled many-particle wavefunctions. Indeed, although the recognition of entanglement is usually attributed to Schrödinger, initially he considered only effective one-particle problems, the hydrogen atom [14], the harmonic oscillator and the rotor [15].

It was Max Born [16] who first treated the two-particle problem of an electron colliding with an atom (to explain the experiment of Franck and Hertz) and wrote down an entangled collision wavefunction. In the same paper Born gave the statistical interpretation of the wave function which is of course the whole basis of the ensemble picture. The theory of entanglement in collisions, involving both the continuous variables of position and momentum and the discrete variables of binding energy, spin and angular momentum has been developed with great sophistication and in countless works in atomic, molecular and nuclear physics, since Born's pioneering paper, see for example [17–20].

The subject of many-particle entanglement has enjoyed enormously renewed interest in the last few years in the fields of quantum information and quantum computing, for example see Ref. [21]. However, this development, often with the limited purview of entangled photon states [22], has been made largely without reference to the study of entanglement in collision physics.

On the basis of the IT, it emerges that whether one ascribes (a) classical dynamics (a single trajectory analogous to a light ray), (b) a quantum wave description of the ensemble as a whole or (c) single particles registered separately whose statistical distribution corresponds to a wave, to the movement of material particles depends upon the precision and extent to which the dynamical variables of position and momentum are determined by the measurement. This is equally true for ensembles of many-particle systems involving entangled wavefunctions as it is for ensembles described by single particle wavefunctions.

The elimination of the overtly quantum effects of entanglement and coherence as a prerequisite for the transition to classical mechanics has been ascribed to interaction with the environment [23–26]. It goes under the broad name of "decoherence theory" (DT). This theory is part of the wider study of open quantum systems and these approaches usually involve propagation of the quantum density matrix in time.
The principal feature of such models is that the interaction with an environment leads to a suppression of the off-diagonal elements of the density matrix, which is considered a key element of the transition to classical behaviour.

Without doubt DT can explain many features of the quantum to classical transition but, according to DT, unitary evolution *in the system Hamiltonian alone* does not contribute to this transition. One main aim of the present work is to show that this is not the case.

Generically, from the result of the IT, a quantum system wavefunction or corresponding density matrix propagating in time without environmental interaction will develop such that the position and momentum coordinates change according to classical mechanics. In particular, the off-diagonal density matrix elements acquire oscillatory phase factors such that, except under a high-resolution measurement, they average to zero. In this sense, the IT does not negate any predictions of DT, rather it is complementary to it. However the unitary propagation transition occurs over time and position increments which are still of atomic dimensions and thus largely obviate any additional changes to the density matrix ascribable to the environment.

An exhaustive discussion of DT with an appraisal of its notable successes but also its limitations is given in the reviews of Schlosshauer [25, 26]. It is clear that this theory is anchored firmly in the SP interpretation of the wavefunction since wavefunction collapse, or apparent collapse, plays a prominent role.

In the present work only continuum quantum states are considered of relevance in the transition to classical mechanics. Bound states and quantised internal degrees of freedom (e.g. intrinsic spin) are viewed as wholly quantum features. They are not affected by the unitary propagation according to the IT of particles as a whole.

However, the quantised angular momentum may be used to affect the classical trajectories [27] exactly as in the original Stern-Gerlach experiment.

Here, there is no discussion of the measurement process itself. In the particle detectors considered in this work the quantum particle is intercepted by a macroscopic detector involving an enormous number of atomic degrees of freedom, giving a completely irreversible transformation. The particle energy is absorbed through ionisation or photon emission in the detector and amplified to give a recorded signal. Such measurements are often called "strong" measurements. Hence, here we do not consider so-called "weak" measurements, usually performed on light [28], which involve additional manipulation of the wavefunction e.g. change of polarisation state, during transit.

2 Interpretation of the Wavefunction

Here a simple but sufficient interpretation of the wavefunction is applied. This involves the minimum of supposition required to explain modern multi-hit coincident detection of particles emanating from complexes of atomic dimension. The following rules are adopted in connection with the detection of moving particles.

(1) The wavefunction always describes a statistical ensemble of identically-prepared particles. No meaning can be ascribed to the wavefunction of a single particle.
(2) The wavefunction $\Psi(\boldsymbol{r}, t)$ contains information on the state of the ensemble. The wavefunction extent can be infinite or spatially confined.
(3) The quantity $|\Psi(\boldsymbol{r}, t)|^2 d\boldsymbol{r}$ gives the probability to detect a given particle from the ensemble at position $\boldsymbol{r}$, at time t and with a resolution $d\boldsymbol{r}$ (Born's rule [16]). The quantity $|\tilde{\Psi}(\boldsymbol{p}, t)|^2 d\boldsymbol{p}$, where $\tilde{\Psi}(\boldsymbol{p}, t)$ is the wavefunction in momentum space, gives the probability to detect a given particle from the ensemble with momentum $\boldsymbol{p}$ at time t.
(4) When information, either partial or total, is extracted by a measurement, the corresponding part of the quantum wavefunction has been utilised and no further information can be extracted from that part.

Consequent on this ensemble view, the popular expression that a particle can also behave as a wave is redundant. What is detected is always a particle. The wavefunction simply assigns a probability amplitude that a particle from an ensemble of identical particles will be detected to have particular values of the dynamical variables.

As will be shown in the following, the IT provides many of the features of wavefunction propagation ascribed to decoherence due to environmental interaction. However, since the propagation is unitary, classical features emerge autonomously.

Wavefunction "collapse" is a widely-accepted aspect of quantum mechanics. This concept is peculiar to the SP picture. In the ensemble picture the need to invoke collapse of the wavefunction does not arise.

3 The Imaging Theorem

The result known as the imaging theorem can be expressed in a few equations. Details of the original proof for free asymptotic motion can be found in the book of Kemble [5] and its generalisation for arbitrary motion, e.g. in external electromagnetic fields, in Ref. [6].

The propagation in time of a localised quantum state defined at time t' can be written

$$|\,\Psi(t)\,\rangle = U(t,t')\,|\,\Psi(t')\,\rangle, \tag{1}$$

where $U(t,t')$ is the time-development operator. Projecting this equation into position space gives

$$\Psi(\boldsymbol{r},t) = \int K(\boldsymbol{r},t;\boldsymbol{r}',t')\,\Psi(\boldsymbol{r}',t')\,d\boldsymbol{r}', \tag{2}$$

where the function $K(\boldsymbol{r},t;\boldsymbol{r}',t') \equiv \langle\,\boldsymbol{r}\,|\,U(t,t')\,|\,\boldsymbol{r}'\,\rangle$ is called the space-time propagator. The IT rests on the asymptotic large r, large t limit when the action becomes much greater than $\hbar$ and the propagator can be approximated by its semi-classical form [29]

$$K(\boldsymbol{r},t;\boldsymbol{r}',0) = \frac{1}{(2\pi i\hbar)^{3/2}}\left|\det\frac{\partial^2 S}{\partial\boldsymbol{r}\partial\boldsymbol{r}'}\right|^{1/2} e^{iS(\boldsymbol{r},t;\boldsymbol{r}',0)/\hbar}, \tag{3}$$

where $S(\boldsymbol{r},t;\boldsymbol{r}',0)$ is the *classical* action function in coordinate space and the initial time t' is taken as the zero of time.

Now it is recognised that the $\boldsymbol{r}'$ integral is confined to a small volume, of atomic dimensions, around $\boldsymbol{r}' \approx 0$, so that the action can be expanded around this point as

$$S(\boldsymbol{r},t;\boldsymbol{r}',0) \approx S(\boldsymbol{r},t;0,0) + \left.\frac{\partial S}{\partial\boldsymbol{r}'}\right|_0 \cdot \boldsymbol{r}'. \tag{4}$$

Then, using the classical relationship $\partial S/\partial\boldsymbol{r}'|_0 \equiv -\boldsymbol{p}$, substitution in the integral Eq. (2) gives a Fourier transform and the result

$$\Psi(\boldsymbol{r},t) \approx (i)^{-3/2}\left(\frac{d\boldsymbol{p}}{d\boldsymbol{r}}\right)^{1/2} e^{iS(\boldsymbol{r},t;0,0)/\hbar}\,\tilde{\Psi}(\boldsymbol{p},0), \tag{5}$$

which is the IT of Kemble, here generalised to arbitrary classical motion.

One notes that the IT rests upon two approximations. The first is the semi-classical approximation of K in Eq. (2). However, in the integral over $\boldsymbol{r}'$, all possible values of $\boldsymbol{r}'$ contribute to the asymptotic wavefunction at $\boldsymbol{r}, t$. It is the recognition that the quantum wavefunction at time zero is limited to a microscopic extent, Eq. (4), that associates a fixed classical momentum $\boldsymbol{p}$ to each final coordinate $\boldsymbol{r}(t)$. That is, each initial $[(\boldsymbol{r}' \approx 0), \boldsymbol{p}]$ value is connected to a fixed final $\boldsymbol{r}, t$ by a classical trajectory. For free motion the connection is simply $\boldsymbol{r} = \boldsymbol{p}\,t/m$, where m is the particle mass.

The essence of the IT result is that the position and momentum coordinates evolve classically but within the shroud of the quantum wave functions. Indeed, one can show [8] that the asymptotic position wave function of Eq. (5) can be viewed as an eigenfunction of both the position and momentum operators. Hence, to the accuracy of the IT, these operators commute and, as emphasised in Ref. [12], there is no obstacle in measuring both momentum and position with arbitrary accuracy.

Then the probability density for detection of a particle of the ensemble is given by

$$|\Psi(\boldsymbol{r}(t))|^2 \approx \frac{d\boldsymbol{p}}{d\boldsymbol{r}}\,|\tilde{\Psi}(\boldsymbol{p},0)|^2. \tag{6}$$

Since the coordinates of the wavefunctions now conform to classical mechanics, this form has a wholly classical, statistical interpretation. An ensemble of particles with probability density $|\tilde{\Psi}(\boldsymbol{p},0)|^2$, defining the probability of occurrence of a certain initial momentum $\boldsymbol{p}$, move on classical trajectories and hence the ensemble members evolve to the position probability density $|\Psi(\boldsymbol{r},t)|^2$.

The factor $d\boldsymbol{p}/d\boldsymbol{r}$ is the *classical* trajectory density of finding the system in the volume element $d\boldsymbol{r}$ given that it started with a momentum $\boldsymbol{p}$ in the volume element $d\boldsymbol{p}$ (see the books by Gutzwiller [29] or Heller [30]). Quantum mechanics provides the initial ensemble momentum distribution located at a microscopic distance $\boldsymbol{r}' \approx 0$. Each element of the initial momentum wave function is then imaged onto the spatial wave function at large distance $\boldsymbol{r}$, where the coordinates are related by classical mechanics.

That is, from Eq. (6) one has the asymptotic equality of probabilities in initial momentum space and final position space, i.e.

$$|\Psi(\boldsymbol{r}(t))|^2\, d\boldsymbol{r} = |\tilde{\Psi}(\boldsymbol{p},0)|^2\, d\boldsymbol{p}. \tag{7}$$

This shows that the loci of points of equal probability of particle detection are classical trajectories. Nevertheless, according to Eq. (5), the wavefunction remains intact.

Clearly, the IT can only be interpreted in the ensemble picture. The wavefunction spreading corresponds to the natural divergence of an ensemble of classical trajectories of differing initial momentum emanating from a microscopic volume and being detected after traversing a macroscopic distance. Nevertheless, estimates of the $\boldsymbol{r}$ and t values at which the semi-classical approximation becomes valid (Ref. [6]) show that this occurs for values which are still microscopic, typically only tens of atomic units, the precise value dependent upon particle masses and energies.

It is to be emphasised that the IT describes classical evolution of the wavefunction variables and the transition to this property arises from unitary quantum propagation i.e. the transition to classical behaviour is autonomous; external interactions are unnecessary. This justifies a routine assumption of experimentalists that one can use classical mechanics to trace a trajectory back from a point on the detector to the quantum reaction zone and is valid even for light particles such as electrons.

The consideration of the quantum to classical transition from a more mathematical viewpoint, so-called semi-classical quantum mechanics, began with the early WKB approximations and Van Vleck's work on time propagators [31]. It was formulated initially for scattering theory, for example by Mott and Massey [17], by Ford and Wheeler [32] and by Brink [33]. A completely general theory emerged later in the work of Berry and Mount [34] and of Miller [35] and Heller [30], for example. Major contributions made by Gutzwiller are to be found in Ref. [29].

In semi-classical scattering theory one examines the transition to a classical cross-section which occurs when the collision energy is much greater than the interaction energies of the collision complex, see for example, [36, 37]. This is to be contrasted with the IT in which quantum systems of atomic dimension are described fully by quantum mechanics but the transition to macroscopic distances by the semi-classical approximation. Then the semi-classical description is valid for all energies, after distances are traversed such that the classical action far exceeds $\hbar$. This is the autonomous aspect of the quantum to classical transition.

4 The Quantum to Classical Transition

4.1 Historical Context

The question of the transition from quantum to classical mechanics in the motion of particles is as old as wave mechanics itself. In the SP picture it is required that in the classical limit the wavefunction of a single particle describes a classical trajectory i.e. a narrow wavepacket. In the ensemble picture the limit is, as embodied in the IT, that the wavefunction describes an ensemble of particles following classical trajectories.

4.2 Schrödinger, Heisenberg and Kennard.

Schrödinger, immediately following his invention of wave mechanics in a sequence of papers in 1926, investigated the classical limit of wave mechanics. In a paper [38] entitled "On the continuous transition from micro- to macro-mechanics" he gave an example of how a packet of waves describing the harmonic oscillator can move in such a way that the displacement of the wavepacket as a whole follows the well-known classical dynamics of the one-dimensional harmonic oscillator. In this calculation Schrödinger repeatedly draws the analogy of superpositions of oscillator eigenfunctions to wavepackets formed from classical normal modes on an oscillating string.

The important point to note here is that Schrödinger was seeking, through the wave equation, to represent a *single particle* as a packet of quantum waves which is so localised in space that it can be perceived as a classical particle. Nevertheless he recognized the limitations of his model, pointing out, for example, that a nondispersive packet can only be built from *bound* eigenfunctions and any admixture of continuum states will result in an expanding wavepacket as in the optical case.

This latter point was taken up by Heisenberg [39] in a lengthy paper on the interpretation of the new quantum mechanics and its relation to classical mechanics. In a section also called "the transition from micro- to macro-mechanics", Heisenberg criticises the relevance of bound states in connection with classical mechanics. To illustrate the difficulty with continuum motion Heisenberg showed that an initial Gaussian wavepacket moving freely will spread in space as a function of time and so cannot represent a single material particle.

A more precise demonstration of the classical aspects of quantum motion can be traced back to 1927 in a paper by Kennard [40]. Kennard showed that the *centroid* of quantum "probability packets" moves according to classical mechanics. In retrospect, Kennard probably deserves recognition for the "Ehrenfest" theorem, but perhaps this is denied him since he couched his proof in the language of matrix mechanics, whereas Ehrenfest [41] used Schrödinger wave mechanics.

Kennard's paper is a very important landmark in the development of the meaning of the wavefunction. Interestingly, this is one of the last papers to utilise predominantly the Born, Heisenberg, Jordan [42] theory of matrix mechanics. Kennard defines a "probability amplitude" $M(q)$ for a variable q in matrix mechanics, which is later shown to be equivalent to the Schrödinger wavefunction $\psi(q)$.

He considers the motion of "probability packets" and shows that, for the cases of free motion or motion in constant electric or magnetic fields, the centroid obeys classical mechanics.

As perhaps the first to emphasise the ensemble picture, Kennard shows that Heisenberg's "proof" of the uncertainty principle is properly formulated as the statistical spread of momentum and position measured on an ensemble of identical systems. The spread, for the particular case of a free wavepacket, is calculated using the probability $MM^* \, dq$ which is identical to Born's probability interpretation of the Schrödinger wavefunction.

As mentioned above, the case of free motion had been solved already by Heisenberg [39] who showed that a Schrödinger free wavepacket spreads in time. Kennard, although he shows that his probability amplitude M is the same as a Schrödinger wavefunction ψ, uses this spreading as an argument against the superiority of Schrödinger wave mechanics with respect to matrix mechanics.

Kennard raises objections to the Schrödinger wave equation by pointing out that a spreading wavefunction of an electron must correspond to a spreading of charge density. Note that here, in contrast to his view of the M of matrix mechanics, in interpreting Schrödinger's ψ, Kennard is assuming that the SP picture applies to this wavefunction. Then he points out that a detection of the electron must localise its full charge at a point. Hence, because of the measurement, the original diffuse wavepacket "loses any further physical meaning" and must be replaced by "a new, smaller wavepacket". Kennard is using the necessity, in the particle picture, to invoke a "collapse of the wavefunction" as an argument against the use of a Schrödinger wavefunction.

Following this objection to the collapse scenario, Kennard then advances the ensemble interpretation of the probability amplitude of matrix mechanics. He writes "the wavepacket spreads, for example, like a charge of shot, in which each pellet describes a trajectory dependent upon its initial position and motion and the whole charge spreads in time as a consequence of differences in these initial conditions", precisely as described by the IT Eq. (6). In the ensemble picture, as distinct from the SP picture, there is no problem with the spreading of the wavepacket. Classical particles with different initial momenta will spread out as they move from micro- to macroscopic distances.

4.3 Ehrenfest and Einstein

Ehrenfest's paper was published a few months after Kennard's. Apparently, the clarification of the connection of quantum to classical mechanics received an enormous boost with this publication. Ehrenfest used the Schrödinger equation to prove the theorem that quantum position and momentum *expectation values* obey a law similar to Newton's law of classical mechanics. In one dimension, using Ehrenfest's notation, it is expressed as

$$m\frac{d^2\langle x\rangle}{dt^2} = \int dx\, \Psi\Psi^* \left(-\frac{\partial V}{\partial x}\right) = -\langle\frac{\partial V}{\partial x}\rangle \tag{8}$$

As often remarked, however, this is not Newton's Law which would require $-\partial\langle V\rangle/\partial x$ to appear on the r.h.s.. However, it turns out that for the cases $V = a$, $V = ax$ and $V = ax^2$, where a is a real constant, the theorem is the same as Newton's law. The spreading of wavepackets remains a problem since, if the wavepacket occupies a macroscopic volume of space, little meaning can be attributed to an average position. Also, for all other potentials with terms higher than quadratic one does not have motion according to Newton's law. Hence, for these two reasons and despite its appealing form, in general Ehrenfest's theorem cannot be considered as describing the transition to classical mechanics, as emphasised by Ballentine [43, 44].

Mindful of Heisenberg's proof of free wavepacket spreading, Ehrenfest is careful to stress that, within the particle picture, the motion of the mean value according to Newtonian mechanics is meaningful only "for a small wavepacket which remains small (mass of the order of 1gm.)". Clearly he was thinking of a single particle described by a small wave packet. The ideas that narrow wavepackets and Ehrenfest's theorem embody the nature of the quantum to classical transition for a single particle, pervade most elementary text books on quantum mechanics even today.

The SP and ensemble pictures were hotly discussed at the Fifth Solvay conference in October 1927. Einstein gave an example of electrons emerging from a small hole to impinge on a distant screen. He pointed out that in the ensemble picture the wave function simply gives the probability of electron detection at a given point. However, in the SP picture of the wave function, the wave which has spread to occupy a macroscopic space, must "collapse" to a point on the screen. Einstein objected to this and commented "one can only remove the objection in this manner, that one does not describe the process by the Schrödinger wave only but at the same time one localises the particle during propagation". Remarkably, after more than ninety years, we recognise that the IT wave function fulfills *exactly the property that Einstein was seeking*, a Schrödinger wave whose variables follow classical trajectories.

4.3.1 After 1927

It is interesting, although understandable in the first years of quantum and wave mechanics, that the SP and ensemble pictures are confused continually. This applies not only to Kennard, as outlined above, but also to Heisenberg and Schrödinger themselves. In discussing the uncertainty principle, Heisenberg describes exclusively measurements on a single particle, as is discussed in great detail in the accompanying paper by Schmidt-Böcking et.al. [12]. This is despite the Kennard paper quoted above and, most importantly, Robertson's proof [13] of the uncertainty principle. Both papers make clear that the spread of measured values of a variable refers to an ensemble statistical spread and not the uncertainty in measuring that property on a single particle. Similarly, Schrödinger, although a confirmed advocate of the SP picture, still admits the validity of Born's statistical interpretation and the necessity to consider a sequence of measurements, see the discussion of Mott's problem given below.

Although the Ehrenfest Theorem and narrow wavepackets are used as the classical limit in many elementary text books, reminders have been given continually since 1927 of the problems involved with this picture and the essential interpretation of a wavefunction as representing an ensemble and not a single particle.

Kemble in 1935 [49], comments that the interpretation of quantum mechanics "asserts that the wavefunctions of Schrödinger theory have meaning primarily as descriptions of the behaviour of (infinite) assemblages of identical systems similarly prepared".

Writing in 1970, Ballentine [43] advances several arguments "in favour of considering the quantum state description to apply only to an ensemble of similarly prepared systems, rather than supposing as is often done, that it exhaustively represents a single physical system". In addition, in 1972 [44] Ballentine, considers "Einstein's interpretation of quantum mechanics" and advances convincing evidence that Einstein was a firm proponent of the ensemble picture. Indeed, Ballentine must be considered as a prophet of the ensemble picture and many of his ideas are corroborated by the arguments advanced in the present paper.

In a scholarly essay in 1980, on the "Probability interpretation of quantum mechanics", Newton [45] emphasises that "the very meaning of probability implies the ensemble interpretation".

In 1994, Ballentine et al. [46] examined the Ehrenfest theorem from the point of view of the quantum/classical transition and concluded that "the conditions for the applicability of Ehrenfest's theorem are neither necessary nor sufficient to define the classical regime." Furthermore, in connection with the ensemble or SP pictures they pointed out that "the classical limit of a quantum state is an ensemble of classical orbits, not a single classical orbit." A comprehensive account of ensemble interpretations of quantum mechanics is given by Home and Whitaker [47].

5 Consequences of the IT and the Ensemble Picture

In this section three classic problems of quantum theory are analysed briefly within the IT and related ensemble picture. The problems are the subject of countless papers and the ensemble aspects have been discussed before. However, the consequences of the IT illuminate further the simplicity of the ensemble explanation. Then the reconciliation of the classical trajectory aspect of IT with the quantum interference effect is presented.

5.1 The Schrödinger Cat

The mere posing of this question by Schrödinger [50] attests to his adherence to the SP interpretation of the wavefunction. As has been observed earlier, in the ensemble picture the interpretation is trivial, as explained by Ballentine [48]. Since the wavefunction applies to many observations, one finds that half the cats are alive and half are dead. No meaning can be attached to the observation of a single cat, unless successive measurements are made over time and feline re-incarnation is allowed.

In the same paper, Schrödinger comments on the apparent problem that radioactive decay described by a spherically-symmetric wave does not lead to uniform illumination of a spherical screen but rather to individual points which slowly are seen to be uniformly distributed. However, although he states that "it is impossible to carry out the experiment with a single radioactive atom" he does not concede that this requires an ensemble interpretation of the wavefunction. This is precisely the problem of Mott which is considered next.

5.2 The "Mott Problem" of Track Structure

One of the oldest "problems" of the interpretation of a wavefunction for material particles is that posed by Schrödinger [50] and addressed in 1929 by Mott [51]. Certainly Mott's paper was at the instigation of his mentor Darwin, a confirmed adherent of the SP picture [52]. This is one of the most striking examples of erroneously assigning a wavefunction to a single particle. Mott remarked,

"In the theory of radioactive disintegration, as presented by Gamow, the α-particle is represented by a spherical wave which slowly leaks out of the nucleus. On the other hand, the α-particle, once emerged, has particle-like properties, the most striking being the ray tracks that it forms in a Wilson cloud chamber. It is a little difficult to picture how it is that an outgoing spherical wave can produce a straight track; we think intuitively that it should ionise atoms at random throughout space."

Mott presents a detailed argument based on scattering theory to argue that only atoms lying on the same straight line will be ionised successively by an α-particle emitted in a spherical wave. Although Mott repeatedly refers to the *probability* of ionisation he interprets the wavefunction as applying to a single α-particle.

However, according to the IT and the ensemble interpretation the proof of Mott is completely superfluous. There is absolutely no mystery attached to "how it is that an outgoing spherical wave can produce a straight track". This apparent dichotomy of wave mechanics is explained by the dual nature of the semi-classical wavefunction of Eq. (5); quantum wavefunction with classically-connected coordinates. Each coordinate of the initial momentum wavefunction corresponds to a specific momentum and therefore to a specific position $\boldsymbol{r}(t)$ along the classical trajectory. The spherical S wavefunction applies to the ensemble as a whole and specifies equal probability of emission in all directions, i.e. uniform distribution of $\boldsymbol{p}$ on the unit sphere. Each α-particle is launched with a certain momentum $\boldsymbol{p}$ distributed according to the initial momentum wavefunction and, according to the IT, the position on a macroscopic cloud chamber scale follows a *straight line* classical trajectory. Hence it is obvious that only atoms lying along this trajectory can be ionised and the usual straight track in the cloud chamber is observed.

This is a prime example of the principle that what one perceives, in this case directed motion (a set of classical trajectories) or a spherically uniform distribution (a quantum S-wave probability) depends upon the nature and duration of the experiment.

5.3 Entanglement and Wavefunction Collapse

That wavefunction superposition applies to an ensemble is made clear also by the process of radioactive decay discussed above. Although usually thought of in the time domain, the stationary picture is simpler. The wave function of an ensemble of nuclei is described by a superposition of the state of a bound nucleus and the state of two separated product nuclei at the same total energy. The intrusion of a measuring device simply detects which state a given member of the ensemble occupies. The absence of a signal in a measuring device denotes undecayed state and a signal denotes a decay. The half-life is interpreted from a sequence of measurements on the ensemble. It is not a property of a single nucleus, although colloquially the half-life is often so ascribed. This aspect is emphasised particularly in the very clear exposition of Rau [53].

The paper of Einstein, Podolsky and Rosen [54], whose result often is referred to as the "EPR paradox", has been the subject of an enormous number of works on the subject of reality, action at a distance etc. Throughout the EPR paper appears the SP viewpoint of a partial wavefunction describing an independent particle.

Already in the first replies to the EPR paper, by Schrödinger [55] and Bohr [56], it was pointed out that it is essential to consider the *two-particle* commuting operators, ignored by EPR. Nevertheless the reply papers did not apply these considerations directly to the EPR entangled wavefunctions.

Here we infer the ensemble picture and show that the recognition of good two-particle quantum numbers is essential. Then, in the pure states considered in EPR, a good quantum number ensures that every pair of the ensemble will give the same value of the corresponding two-particle property upon measurement.

EPR consider a two-particle eigenstate written in the entangled form

$$\Psi(x_1, x_2) = \int \psi_p(x_2) u_p(x_1)\, dp \tag{9}$$

where

$$u_p(x_1) = e^{\frac{i}{\hbar} p x_1} \quad \text{and} \quad \psi_p(x_2) = e^{-\frac{i}{\hbar} p (x_2 - x_0)} \tag{10}$$

are eigenfunctions of one-particle operators p_1, p_2 with eigenvalues p and $-p$ respectively. The constant x_0 is arbitrary. Note that the single-particle momentum p can take any value.

The p integral in this equation can be carried out to give

$$\begin{aligned} \Psi(x_1, x_2) &= 2\pi \delta(x_1 - x_2 + x_0) \\ &= 2\pi \int \delta(x_1 - x)\delta(x - x_2 + x_0)\, dx \end{aligned} \tag{11}$$

which is an entangled state in position space. However, again, all x values are possible.

Thus it has been shown that one and the same two-particle function can be expanded in terms of eigenfunctions of observables of particle 2, in this case p and x, which do not commute.

As shown by Schrödinger [55] and Bohr [56], the conserved quantities emerge from a transformation to relative and centre-of-mass (CM) coordinates for equal mass m particles. We define relative x_r and CM position X as

$$x_r = x_1 - x_2 \quad \text{and} \quad X = (x_1 + x_2)/2 \tag{12}$$

and correspondingly relative and CM momenta

$$p_r = (p_1 - p_2)/2 \quad \text{and} \quad P_{CM} = p_1 + p_2 \tag{13}$$

Immediately one sees from Eq. (11), that $\Psi(x_1, x_2)$ *is* an eigenfunction of the relative position coordinate $x_r = x_1 - x_2$ with eigenvalue $x_r = -x_0$. Similarly, from Eq. (9) with Eq. (10) one sees it is simultaneously an eigenfunction of CM momentum P_{CM} with eigenvalue zero. This is in order since these two operators commute. However it is readily checked, as must be, that $\Psi(x_1, x_2)$ *is not* an eigenfunction of X or p_r since these do not commute with P_{CM} and x_r respectively.

In summary, the two-particle wavefunction of EPR fixes the CM momentum at zero and the relative position of the particle pair is equal to $-x_0$. This is the only information in the two-particle wavefunction. One has, however, the clear requirement that the two-particle wavefunction should propagate intact to the detectors. In any measurement the two corresponding two-particle observables have the same precise value for all members of the ensemble of pairs.

Now one has two possible scenarios characterising entanglement.

(a) If one knows the two-particle good quantum numbers in advance e.g. by selection rules on state preparation, then the determination of the single-particle momentum to be $p = p_1$ fixes $p_2 = -p_1$. Similarly measurement of x_1 fixes $x_2 = x_1 + x_0$.
(b) If one does not know the quantum numbers in advance, one must perform measurements on many two-particle systems *in coincidence*. Then one can ascertain by experiment that, for all ensemble members, whatever the measured values of p_1 and x_1, one measures always $p_2 = -p_1$ and $x_2 = x_1 + x_0$.

The measured two-particle eigenvalues are sharp, $P_{CM} = p_1 + p_2 = 0$ and $x_r = x_1 - x_2 = x_0$ for all members of the ensemble with no statistical spread, in accordance with their commutation. Note that the specification of *two-particle* conserved observables allows one to assign precise values to both non-commuting *one-particle* observables. Hence there is absolutely no barrier to measuring both position and momentum of one or both of the particles with arbitrary accuracy. This is emphasised by the analysis of the Uncertainty Relation in the accompanying paper of Schmidt-Böcking et.al. [12]. Of course the *single-particle p* values have a distribution of probability predicted by projection of the one-particle probability amplitude out of the two-particle wavefunction.

Both scenarios require non-local information. The measurement in (b) requires communication between the two separated detectors to ensure coincidence. In case (a) only one detector is required but the non-local information is in the knowledge of the two-particle quantum numbers which are conserved for all particle separations.

The simultaneous fixing of position and momentum becomes apparent within the IT if, as is normal, detection is made at large distances from the volume from which the correlated pair is created. According to the IT there is a classical connection between position and initial momentum for detection of particles 1 and 2 at times t_1 and t_2 respectively. Then the space wavefunction can be written

$$\Psi(x_1, x_2) \propto \tilde{\Psi}(p_1, p_2 = -p_1). \tag{14}$$

In particular the IT gives the classical relation

$$x_1 = p_1 t_1/m \text{ and } x_2 = -p_2 t_2/m \tag{15}$$

so that from the second conservation law $x_2 = x_1 + x_0$ one has the restriction

$$x_0 = -(p_1 t_1 + p_2 t_2)/m. \tag{16}$$

Single-particle x and p can be measured simultaneously with sub-$\hbar$ accuracy, see Ref. [12].

A striking manifestation of such entanglement, which has been well-studied in experiments, is the full fragmentation of the helium atom by a single photon. This example is given since it comprises simultaneously both the momentum and position (continuous variable) entanglement of EPR *and the discrete variable spin entanglement*, as envisaged by Bohm [2], in a pure two-electron state. Furthermore, from the IT, the electrons can be assigned classical trajectories *within the two-electron quantum wavefunction.* This is not a "Gedankenexperiment" but a real measured system [57].

The two electrons emerging can be detected in coincidence and occupy a $^1P^o$ two-electron continuum state (this means their state is a spin singlet, has total orbital angular momentum one unit and odd parity). A selection rule [58] says that electrons of the same energy cannot be ejected back-to-back i.e at 180° such that $\boldsymbol{p}_1 = -\boldsymbol{p}_2$. That is, the *two-electron* state has a node for the EPR configuration as the coincidence experiments confirm.

If one of the electrons is left undetected a counter will register electrons of a given energy at a particular angle. However, if a detector diametrically opposed is switched on to detect electrons of the same energy in coincidence, the counts in both detectors will be zero. This coherent state can be made incoherent by switching off one of the detectors when electrons will be measured again. The essence is that this pure effect of wavefunction entanglement is evident, even though according to the IT, the electrons are moving on classical trajectories after they exit the reaction zone with well-defined momenta.

In interpreting the wavefunction, as in EPR, it is crucial that the ensemble is viewed as an ensemble of *two-electron* systems. This two-electron wavefunction is the single quantum entity and it must be transmitted to the macroscopic detection zone unchanged. Then there is no wavefunction interpretation problem within the ensemble picture. The wavefunction node says that there is zero probability that a given member of the ensemble (a coincident pair of electrons) will be emitted in the forbidden configuration.

The coincident detection of position and momentum extracts the information from the wavefunction of the ensemble of two-electron states. The non-coincident detection of electrons extracts information only on the ensemble of single electrons. The effect of entanglement is non-local simply because the two-electron wavefunction is non-local.

A comprehensive discussion of the implications of entanglement for the uncertainties in measured properties relevant to quantum information in the case of the continuous variable description of light rather than material particles, is given by Braunstein and van Loock [22].

5.4 Quantum Interference

The Davisson-Germer experiment of 1927 on electron-beam diffraction established the validity of the description of particle ensembles by a wave function. The diffraction of heavy neutral particle beams was confirmed in the pioneering experiments of Stern and co-workers as early as 1929 [59]. The demonstration of interference even of large molecules has been achieved recently in the remarkable experiments of Arndt and his group [60], reported in Chap. 24 of this volume.

The explanation of interference patterns in terms of semi-classical wavefunctions and the underlying classical trajectories has been given in great detail by Kleber and co-workers [61] and will not be repeated here. Based upon the IT (see eq.(1) of Kleber [62]), their theory is used to interpret experiments such as those of Blondel et.al. [63]. Here the "photoionisation microscope" exhibits interference rings of electrons ionised from a negative ion in the presence of an extracting electric field. In the semi-classical explanation electrons can occupy two classical trajectories. Either they proceed directly to the detector or, initially they are ejected moving away from the detector but are turned around in the electric field. The imaging of the spatial wavefunction squared is obtained by detection on a fixed flat screen i.e. only the position of electrons is detected. Then an interference pattern from the two trajectories is observed.
However, were the vector position *and* vector momentum of the electrons to be observed, that would correspond to a "which way" determination and the perception would be of two distinct non-interfering classical trajectories. Interestingly, as distinct from entanglement, in this case it is a lack of information which gives rise to wave perception. Blondel et al. [63] remark also that for ionisation from neutral atoms the interference rings are there but are too small to be detected, again showing that perception depends upon resolution.

6 The Imaging Theorem and Decoherence Theory: IT and DT

As stated in the Introduction, the suppression of state superposition, entanglement and interference through environmental interaction can be seen as a requirement on the way to a classical limit of quantum mechanics and has come to be known as "decoherence theory" (DT). It is viewed as a universal phenomenon, extending even to the classical limit of quantum gravity [64, 65] (for an interesting discussion see Ref. [66]). In the following the transition from quantum to classical perception is discussed.

6.1 Decoherence

There is an enormous literature on DT and associated theories describing "spontaneous localisation" due to stochastic interaction. Space does not permit a discussion of the many and varied aspects of these theories, so here consideration is given to those features relevant to the quantum to classical transition embodied in the IT and to the SP or ensemble interpretation of the wavefunction.

The essence of DT is given in the famous paper of Zurek [24] and in more detail in the reviews of Schlosshauer [25, 26]. A more exhaustive treatment with discussion of the $\hbar$ dependence of the environmental interaction terms is to be found in the stochastic Schrödinger equation approach [67]. Here the simpler original density matrix version of Zurek [24] is sufficient as illustration.

The basic mechanism of DT by which certain quantum aspects are eliminated is quite straightforward, accounting for the universality of this phenomenon. In the simplest case presented in Ref. [25], a one-dimensional two-state quantum system S, with states $|\, \psi_n \,\rangle$, is assumed to become entangled with an "environment" with corresponding states $|\, E_n \,\rangle$. Limiting to two-state quantum systems, the ensuing entangled state vector is

$$|\, \Psi \,\rangle = \alpha |\, \psi_1 \,\rangle |\, E_1 \,\rangle + \beta |\, \psi_2 \,\rangle |\, E_2 \,\rangle \tag{17}$$

and gives a total density matrix $\rho = |\, \Psi \,\rangle\langle\, \Psi \,|$. According to Ref. [25], "the statistics of all possible local measurements on S are exhaustively encoded in the reduced density matrix ρ_S", given by

$$\begin{aligned} \rho_S &= Tr_E \rho \ = |\alpha|^2 |\, \psi_1 \,\rangle\langle\, \psi_1 \,| + |\beta|^2 |\, \psi_2 \,\rangle\langle\, \psi_2 \,| \\ &\quad + \alpha\beta^* |\, \psi_1 \,\rangle\langle\, \psi_2 \,|\langle\, E_2 \,|\, E_1 \,\rangle + \alpha^*\beta |\, \psi_2 \,\rangle\langle\, \psi_1 \,|\langle\, E_1 \,|\, E_2 \,\rangle. \end{aligned} \tag{18}$$

Then a measurement of the particle's position is given by the diagonal element,

$$\begin{aligned} \rho_S(x,x) =& |\alpha|^2 \, |\psi_1(x)|^2 + |\beta|^2 \, |\psi_2(x)|^2 \\ &+ 2\mathrm{Re}[\alpha\beta^* \psi_1(x)\psi_2^*(x)\langle\, E_2 \,|\, E_1 \,\rangle] \end{aligned} \tag{19}$$

where "the last term represents the interference contribution". The assumption of DT is that in general the states of the environment are orthogonal and so the interference term disappears. More importantly, from Eq. (18) the off-diagonal terms disappear and one has a diagonal density matrix only. From Eq. (19) this has two "classical" terms interpreted as classical probabilities.

A slightly different model is adopted in Ref. [24] in that the two states comprising the system S are taken as two spatially-separated Gaussian wavefunctions. The corresponding system density matrix exhibits four peaks. This density matrix is propagated in time subject to a temperature-dependent environment interaction. The result is to give a density matrix of diagonal form with only two peaks along the diagonal.

In this case the decoherence reduces the off-diagonal elements to zero and the diagonal term does not contain the "interference" contribution since the Gaussians do not overlap. This removal of coherence between different spatial parts of the wavefunction is considered to correspond to the emergence of classicality. In connection with the classical transition Schlosshauer writes [26] "the interaction between a macroscopic system and its environment will typically lead to a rapid approximate diagonalisation of the reduced density matrix in position space and thus to spatially localised wavepackets that follow (approximately) Hamiltonian trajectories". This following of classical trajectories however, is not proven in detail.

Implicit is the SP picture in which the diagonal elements represent narrow wavepackets giving classical behaviour via Ehrenfest's theorem. The ultimate spreading of these wavepackets is not considered, although suitable environmental interaction can lead to the wavepackets remaining narrow. In short, the transition to classicality is viewed as an elimination of quantum coherence effects and the vital feature of the emergence of classical dynamics according to Newton is not shown.

6.2 *Unitary Evolution*

In appendix A, following the example of Ref. [24], the free *unitary* propagation of two, initially narrow, Gaussian wavepackets within the IT is calculated. It is shown that, under low detector resolution, the density matrix also assumes the diagonal form

$$\rho(x, x, t) = \frac{1}{\sqrt{\pi}\,\eta(t)}\,(e^{-(x-X_1)^2/\eta^2} + e^{-(x-X_2)^2/\eta^2}) \tag{20}$$

where X_1, X_2 are the centres of the wavepackets and the time-dependent width is $\eta = \tilde{\sigma}t/\mu$, for initial width $\tilde{\sigma}$ and particle mass μ. Hence, the intrinsic spreading of the wavepacket with time emerges as expected in the ensemble picture. In this picture there is no problem of interpretation of the two probabilities; 50% of the ensemble members will be detected near to X_1 and 50% near to X_2. Wavefunction collapse is unnecessary. Most important however, in the IT, the propagation of the co-ordinates of the diagonal density matrix is according to *classical* mechanics. Nevertheless, if the resolution is on the microscopic scale then interference and manifestations of quantum propagation resulting from finite off-diagonal elements can be detected. Just as in optics, the perception of particle trajectory (ray) or wave is decided by the sharpness of vision.

The study of collision complexes in nuclear, atomic and molecular physics has long been concerned with the questions of measurement of interference and entanglement effects [18–20]. Coincidence detection of several collision fragments in entangled states are performed with increasing sophistication (see, for example, Ref. [11]). In line with the IT, classical motion of the collision fragments outside the reaction zone is shown to be appropriate. Nevertheless quantum coherence is preserved showing that environmental decoherence does not occur in such experiments.

The degree of decoherence assigned to a many-body entangled state depends upon which particles are *not* observed or even which dynamical properties are observed and which are not. Coherence can be fully or partially removed according to the experiment. In the language of the experimentalist, either one registers the "coincidence" spectrum or the "singles" spectrum. Again this illustrates that perception of quantum effects depends upon the measurement. Non-detection of collision variables corresponds to a partial trace of the full density matrix, as in DT.

7 Conclusions

The imaging theorem corresponds only to the ensemble interpretation. According to the IT, an initial momentum wavefunction decides the spatial wavefunction at macroscopic distance. The modulus squared of the initial momentum wavefunction corresponds to an ensemble of classical particles with the same initial momentum distribution. Each particle appears to move along a classical trajectory to be registered at well-defined position at a distant screen. Nevertheless the quantum wavefunction is preserved so that the loci of points of equal probability are the classical trajectories but that probability is given by the quantum position wavefunction.

Indeed, all collision experiments support the ensemble picture. One counts many particles at different locations and times on a detector and so builds an image of the initial momentum distribution. Particularly striking in this respect is the observation of the gradual assembly of an interference pattern. Using electron diffraction through a pair of slits, it has been shown [4] that the wave interference pattern is built up slowly by registering many hundreds of hits of individual electrons on a detector screen. Even more strikingly, the experiment has been performed with very large organic molecules [69]. This shows convincingly that it is the ensemble of hits at the detector that builds up the wave interference pattern.

In contrast, the SP picture is that the wavefunction, extending over macroscopic distance, is carried by each molecule and the wavefunction collapses at different points on the screen. That is, the detector is required to instigate decoherence leading to instantaneous wavefunction collapse (from macroscopic to microscopic extent) and the electron being registered at a single localised point on the detector. Again, one is faced with the dilemma of Kennard and Einstein in accepting the plausibility of such a transition.

To summarise, it has been shown that;

(1) The IT preserves the quantum wavefunction out to macroscopic distances but the momentum and position *coordinates* change in time according to classical mechanics. With the IT asymptotic wave function, position and momentum can be measured with arbitrary accuracy [12].
(2) As a result of the IT, unitary evolution of quantum systems, even after propagation over relatively microscopic distances, leads to perception of an ensemble of particles as following classical trajectories.

(3) Standard measurement techniques, either on single or multiple particles, can lead to perception or otherwise of the quantum properties of interference and entanglement according to the information registered. The inference of classical or quantum behaviour depends ultimately upon the resolution and detail of the measurement performed.

Without environment influence, within the IT, unitary evolution of quantum systems results in effective decohering effects. This "decoherence" is of a different nature than in DT. It occurs due to cancellation of oscillating terms of different phase, which leads to non-resolution of oscillatory terms in the propagation of the density matrix to macroscopic times, as in Eq. (27) and Eq. (28). Hence, lack of sufficient resolution results in effective decoherence although paradoxically it arises from the very terms, oscillatory phase factors, which are the hallmark of quantum coherence in the wavefunction.

Already in 1951, long before the formulation of DT, in his discussion of the motion of atom beams in the Stern-Gerlach experiment [1], Bohm [2] points out the decoherence arising from interaction with a macroscopic detector. Interestingly, he attributes the lack of interference also to the impossibility of resolving oscillatory energy phase factors in the unitary time propagation (exactly as in the Appendix) but does not really emphasise the distinction between this and the DT interactions.

The preservation of the wavefunction can lead to interference. However, the perception of interference patterns, or not, again depends upon the nature of the measurement performed. The observation of interference patterns implies that, although resolution is high, incomplete information as to the different trajectories encoded in the wavefunction variables is extracted by the measurement. That is, a "which way" detection is not performed. Then, whether one perceives quantum or classical dynamics depends simply upon the precision of the measurement performed and the amount of information extracted from the wavefunction. This is all in close analogy, both physically and mathematically, to the optical case of perception of particle, wave or ray properties.

In the case of the detection of the effects of particle entanglement it is necessary to treat the ensemble entity as corresponding to the *many-particle* wavefunction and its quantum numbers. Incomplete extraction of the information encoded in the many-particle ensemble wavefunction, for example detection of only some of the particles or incomplete specification of vector variables, corresponds to an effective decoherence.

Acknowledgements The author is extremely grateful to Prof. James M. Feagin for several years of close cooperation on the derivation and meaning of the imaging theorem. The insight of Prof. Horst Schmidt-Böcking on experimental methods and their significance for the uncertainty relation, described in the accompanying paper (Chap. 12 of this volume), is acknowledged also.

8 Appendix

As in the discussion of decoherence by Zurek [24] the time development of a one-dimensional single-particle ensemble wavepacket is considered. The wavepacket is composed of two Gaussians centred at $x = X_1$ and $x = X_2$ with width such that there is essentially no overlap at $t = 0$. The initial state is then

$$\Psi(x, t = 0) = (\pi\sigma^2)^{-1/4} \sum_{i=1,2} e^{-x_i^2/(2\sigma^2)} \tag{21}$$

where $x_i \equiv x - X_i$. For $t > t_0$ this initial wavefunction propagates freely in time and has the exact form

$$\begin{aligned}\Psi(x, t) = (\sigma^2/\pi)^{1/4} \left(\sigma^2 + \frac{i\hbar t}{\mu}\right)^{-1/2} \\ \times \sum_{i=1,2} \exp\left[-\frac{x_i^2}{2\left(\sigma^2 + \frac{i\hbar t}{\mu}\right)}\right],\end{aligned} \tag{22}$$

where μ is the particle mass. The IT condition emerges in the limit of large times and distances. Large times corresponds to $\hbar t/\mu \gg \sigma^2$. Then the spatial wavefunction assumes the IT form,

$$\begin{aligned}\Psi(x, t) \approx \left(\frac{\sigma^2}{\pi}\right)^{1/4} \left(\frac{\mu}{i\hbar t}\right)^{1/2} \\ \times \sum_{i=1,2} e^{-(\mu x_i \sigma/(\sqrt{2}\hbar t))^2} e^{i\mu x_i^2/(2\hbar t)}\end{aligned} \tag{23}$$

The IT limit giving the classical trajectory is such that x_i and t both become large but the ratio is a constant classical velocity. To emphasise this we introduce the momenta $p_i = \mu x_i/t$. We also define, as the width of the Gaussian in momentum space, $\tilde{\sigma} \equiv \hbar/\sigma$. Then we can simplify the asymptotic spatial wavefunction using

$$(\mu x_i \sigma/(\sqrt{2}\hbar t))^2 \equiv p_i^2/(2\tilde{\sigma}^2) \tag{24}$$

and the energy phases

$$\mu x_i^2/(2\hbar t) = p_i^2 t/(2\mu\hbar). \tag{25}$$

The asymptotic spatial wavefunction is then,

$$\Psi(x, t) \approx \left(\frac{\mu}{i\sqrt{\pi}\tilde{\sigma} t}\right)^{1/2} \sum_{i=1,2} e^{-p_i^2/(2\tilde{\sigma}^2) + ip_i^2 t/(2\mu\hbar)} \tag{26}$$

which looks exactly like a pair of free momentum Gaussians propagating in time and corresponds to the 1D form of the IT of Eq. (5), with $dp_i/dx_i = \mu/t$ for free motion.

The diagonal element of the density matrix is defined as $\rho(x, x, t) = \Psi^*(x, t)\Psi(x, t)$ and is

$$\rho(x, x, t) = \frac{\mu}{\sqrt{\pi}\tilde{\sigma}t} \sum_{i=1,2} e^{-p_i^2/\tilde{\sigma}^2} + 2\cos\left[(p_1^2 - p_2^2)t/(2\mu\hbar)\right] e^{-(p_1^2+p_2^2)/(2\tilde{\sigma}^2)} \tag{27}$$

The off-diagonal density matrix is defined as $\rho(x, x', t) = \Psi^*(x, t)\Psi(x', t)$ and consists of four terms,

$$\rho(x, x', t) = \frac{\mu}{\sqrt{\pi}\tilde{\sigma}t} \sum_{i,j=1,2} e^{-(p_i^2+p_j'^2)/(2\tilde{\sigma}^2)} e^{-i(p_i^2-p_j'^2)t/(2\mu\hbar)} \tag{28}$$

At $t = 0$ this gives rise to four Gaussian peaks, as in Ref.[24]. It reduces to the diagonal element when $p_i = p_i'$, i.e. $x = x'$ as it should.

One sees that the diagonal matrix element shows two peaks at $p_1 = 0$, $p_2 = 0$ or equivalently $x = X_1$, $x = X_2$. There is also an interference term. In the off-diagonal element there are four peaks, with the two additional peaks at $x' = X_1$ and $x' = X_2$. These also contain oscillatory phase factors giving interference.

Clearly, to observe interference effects the temporal resolution must typically be less than one oscillation, i.e. $t < 4\mu\pi/(p_1^2 - p_2^2)$. Consider that the particles are electrons with mass unity in atomic units (a.u.). If we take the two peaks to be separated by 1 a.u. of distance, then we have $t < 4\pi \approx 10^{-16}$ s. However, typical resolutions are nanoseconds, that is seven orders of magnitude larger than this. If the resolution is $\delta t \equiv \tau$ then the measurement must be integrated over this time period. Typically the oscillatory terms will then give, omitting constants

$$\int_{-\tau/2}^{\tau/2} e^{i(p_1^2-p_2^2)t}\, dt \approx \delta(p_1^2 - p_2^2). \tag{29}$$

and similarly for the off-diagonal element when p_i is replaced by p_i'. In other words, the oscillations will average to zero under low resolution of measurement on an atomic time scale. From Eq. 27 this implies that the density matrix will exhibit only two diagonal gaussian peaks for such measurements,

$$\rho(x, x, t) = \frac{\mu}{\sqrt{\pi}\tilde{\sigma}t}\left(e^{-p_1^2/\tilde{\sigma}^2} + e^{-p_1^2/\tilde{\sigma}^2}\right) \tag{30}$$

with $p_i = \mu(x - X_i)/t$. For the off-diagonal elements, from Eq. (28), all the terms will average to zero under normal time resolution to give zero off-diagonal elements. This is exactly the limit, elimination of off-diagonal density matrix elements, given by

Zurek [24] as the classical limit and resulting from time propagation in the presence of an interacting environment. However, we emphasise again that the wavepackets on the diagonal are spreading and only in the limit that particles are macroscopically massive can this be ignored to give localised single particles as envisaged in [25].

By contrast the IT proves that classicality emerges from unitary Hamiltonian propagation under low temporal resolution, in that the density matrix then has only two diagonal peaks . Quantum coherence is lost except where the temporal and spatial resolution are extremely high. The peaks represent an ensemble of classical particles spreading on classical trajectories and distributed according to the initial Gaussian momentum wavefunction.

References

1. W. Gerlach, O. Stern Zeits. f. Phys. **9**, 349 (1922)
2. D. Bohm, *Quantum Theory* (Prentice-Hall, New York, 1951)
3. K. Gottfried, T.-M. Yan, *Quantum Mechanics: Fundamentals*, 2nd ed. (Springer, New York, 2003)
4. R. Bach, D. Pope, S.-H. Liou, H. Batelaan, New. J. Phys. **15**, 033018 (2013)
5. E.C. Kemble, *Fundamental Principles of Quantum Mechanics with Elementary Applications* (McGraw Hill, 1937)
6. J.S. Briggs, J.M. Feagin, New J. Phys. **18**, 033028 (2016)
7. J.S. Briggs, J.M. Feagin, J. Phys. B: At. Mol. Opt. Phys. **46**, 025202 (2013)
8. J.M. Feagin, J.S. Briggs, J. Phys. B: At. Mol. Opt. Phys. **47**, 115202 (2014)
9. M.R.H. Rudge, M.J. Seaton, Proc. Roy. Soc. London, A **283**, 262 (1965); E.A. Solovev, Phys. Rev. A **42**, 1331 (1990); T.P. Grozdanov, E.A. Solovev, Eur. Phys. J. D **6**, 13 (1999); M. Kleber, Phys. Rep. **236**, 331 (1994); V. Allori, D. Dürr, S. Goldstein, N. Zanghí, J. Opt. B: Quantum Semiclass. Opt. **4**, S482 (2002); M. Daumer, D. Dürr, S. Goldstein, N. Zanghí, J. Stat. Phys. **88**, 967 (1997)
10. J.H. Macek in *Dynamical Processes in Atomic and Molecular Physics*, ed. by G. Ogurtsov and D. Dowek (Bentham Science Publishers, ebook.com, 2012)
11. J. Ullrich, R. Moshammer, A. Dorn, R. Doerner, L. P. H. Schmidt, and H. Schmidt-Boecking, Rep. Prog. Phys. 66, 1463 (2003), M. Gisselbrecht, A. Huetz, M. Lavolle, T. J. Reddish, and D. P. Seccombe, Rev. Sci. Instr. **76**, 013105 (2013), P. C. Fechner and H. Helm, Phys. Chem. Chem. Phys. **16**, 453 (2014)
12. H. Schmidt-Böcking, S. Eckart, H.J. Lüdde, G. Gruber, T. Jahnke Chapter 12 of this volume
13. H.P. Robertson, Phys. Rev. **34**, 163 (1929)
14. E. Schrödinger, Ann. der Phys. **79**, 361 (1926)
15. E. Schrödinger, Ann. der Phys. **79**, 489 (1926)
16. M. Born, Zeits. Phys. **38**, 803 (1926)
17. N.F. Mott, H.S.W. Massey, *Theory of Atomic Collisions* (OUP, Oxford, 1965)
18. U. Becker, A. Crowe (eds.), *Complete Scattering Experiments* (Kluwer Academic, New York, 2001)
19. U. Fano, J.H. Macek, Rev. Mod. Phys. **45**, 553 (1973)
20. K. Blum, *Density Matrix Theory and Applications* (Plenum, New York, 1981)
21. R. Horodecki, P. Horodecki, M. Horodecki, K. Horodecki, Revs. Mod. Phys. **81**, 865 (2009)
22. S.L. Braunstein, P. van Loock, Revs. Mod. Phys. **77**, 513 (2005)
23. From the very extensive literature, see for example E. Joos, H. D. Zeh, C. Kiefer, D. Guilini, J. Kupsch, I.-O. Stamatescu (eds.), *Decoherence and the Appearance of a Classical World in Quantum Theory*, 2nd Ed. (Springer, New York, 2003) and references therein, W.H. Zurek, Phys. Today **67**, 44 (2014) and Los Alamos Science, Number 27 (2002) and references therein, J.J. Halliwell, Phys. Rev. D **39**, 2912 (1989)

24. W.H. Zurek, Physics Today October (1991)
25. M. Schlosshauer, Rev. Mod. Phys. **76**, 1267–1305 (2004)
26. M. Schlosshauer in M. Aspelmeyer, T. Calarco, J. Eisert, F. Schmidt-Kaler (eds.), *Handbook of Quantum Information* (Springer, Berlin/Heidelberg, 2014)
27. M. Brouard, D.H. Parker, S.Y.T. van de Meerakker, Chem. Soc. Rev. **43**, 7279 (2014)
28. S. Kocsis et al., Science **332**, 1170 (2011)
29. M.C. Gutzwiller, *Chaos in Classical and Quantum Mechanics*, 2nd edn. (Springer, New York, 1990)
30. E.J. Heller, *The Semiclassical Way* (Princeton U.P., Princeton and Oxford, 2018)
31. J.H. Van Vleck, P.N.A.S. **14**, 178 (1928)
32. K.W. Ford, J.A. Wheeler, Ann. Phys. (NY) **7**, 259 (1959)
33. D.M. Brink *Semi-classical Methods for Nucleus-Nucleus Scattering* (CUP, Cambridge, New York, 1985)
34. M.V. Berry, K.E. Mount, Rep. Prog. Phys. **35**, 315 (1972)
35. W.H. Miller, Acct. Chem. Res. **4**, 161 (1971)
36. J.M. Rost, E.J. Heller, J. Phys. B **27**, 1387 (1994)
37. J.M. Rost, Phys. Rep. **297**, 271 (1998)
38. E. Schrödinger, Die Naturwissenschaften **14**, 664 (1926)
39. W. Heisenberg, Zeit. f. Phys. **43**, 172 (1927)
40. E.H. Kennard, Zeit. f. Phys. **44**, 326 (1927)
41. P. Ehrenfest, Zeit. f. Phys. **45**, 455 (1927)
42. M. Born, W. Heisenberg, P. Jordan, Zeit. f. Phys. **35**, 557 (1926)
43. L.E. Ballentine, Revs. Mod. Phys. **42**, 358 (1970)
44. L.E. Ballentine, Am. J. Phys. **40**, 1763 (1972)
45. R.G. Newton, Am. J. Phys. **48**, 1029 (1980)
46. L.E. Ballentine, Am. J. Phys. **55**, 785 (1987)
47. D. Home, M.A.B. Whitaker, Phys. Reports **210**, 223 (1992)
48. L.E. Ballentine, Ensembles in Quantum Mechanics in *Compendium of Quantum Physics*, ed. by D. Greenberger, K. Hentschel, F. Weinert (Springer, Berlin, 2009), p. 199
49. E.C. Kemble, Phys. Rev. **47**, 973 (1935)
50. E. Schrödinger, Die Naturwissenschaftern **48**, 807 (1935)
51. N.F. Mott, Proc. Roy. Soc. London, A **126**, 79 (1929)
52. C.G. Darwin, Proc. Roy. Soc. London, A **l17**, 258 (1927)
53. A.R.P. Rau, Phys. Essays **30**, 60 (2017)
54. A. Einstein, B. Podolsky, N. Rosen, Phys. Rev. **47**, 777 (1935)
55. E. Schrödinger, Proc. Cam. Phil. Soc. **31**, 555 (1935)
56. N. Bohr, Phys. Rev. **48**, 555 (1935)
57. J.S Briggs, V. Schmidt, J. Phys. B **33**, R1 (2000)
58. F. Maulbetsch, J.S. Briggs, J. Phys. B **26**, 1679 (1994)
59. I. Estermann, O. Stern, Zeits. f. Physik **61**, 95 (1930)
60. F. Kialka et al., Phys. Scr. **94**, 034001 (2019)
61. T. Kramer, C. Bracher, M. Kleber, J. Phys. A **35**, 8361 (2002)
62. M. Kleber, Phys. Rep. **236**, 331 (1994)
63. C. Blondel, C. Delsart, F. Dulieu, Phys. Rev. Lett. **77**, 3755 (1996)
64. J.J. Halliwell, Phys. Rev. **D39**, 2912 (1989)
65. C. Kiefer in E. Joos, H.D. Zeh, C. Kiefer, D. Guilini, J. Kupsch, I.-O. Stamatescu (eds.) *Decoherence and the Appearance of a Classical World in Quantum Theory*, 2nd Ed. (Springer, New York, 2003), p. 181
66. S.E. Rugh, H. Zinkernagel in K. Chamcham, J. Silk, J. Barrow and S.Saunders (eds.) *The Philosophy of Cosmology* (CUP, Cambridge, 2016)
67. L. Diósi, W.T. Strunz, Phys. Lett. A **235**, 569 (1997)
68. D. Akoury et al., Science **318**, 949 (2007)
69. T. Juffmann et al., Nature Nano **7**, 297 (2012)

7

Testing Fundamental Physics by using Levitated Mechanical Systems

Hendrik Ulbricht

Abstract We will describe recent progress of experiments towards realising large-mass single particle experiments to test fundamental physics theories such as quantum mechanics and gravity, but also specific candidates of Dark Matter and Dark Energy. We will highlight the connection to the work started by Otto Stern as levitated mechanics experiments are about controlling the centre of mass motion of massive particles and using the same to investigate physical effects. This chapter originated from the foundations of physics session of the Otto Stern Fest at Frankfurt am Main in 2019, so we will also share a view on the Stern Gerlach experiment and how it related to tests of the principle of quantum superposition.

1 Introductory Remarks

Experimentally, this research programme is about gas-phase experiments with large-mass particles, large compared to the mass of a single hydrogen atom, in order to test fundamental theories without the influence of the environment, which typically results in coherence-spoiling noise and decoherence effect. Tests of fundamental theories, such as quantum mechanics and gravity, are in the low-energy regime of non-relativistic velocities and therefore far away from a parameter regime of high-energy particle physics considerations. Fundamental theories will be tested in a new regime.

While Otto Stern's pioneering experiments [1], aligned with a fantastically bold and clear research programme, were about the study and control of freely propagating atoms and molecules in particle beams, we here make use of optical, magnetic and electric fields to trap and manipulate single particles, consisting of many atoms, in order to study the new physics and chemistry. The challenge here is to have a strong enough handle on the motion of the particle. For instance, the optical dipole force $F = \alpha \nabla E^2$ is strong and able to trap individual atoms and atomic ensembles making use of resonance effects. This is impossible for large molecules, again such

H. Ulbricht (✉)
School of Physics and Astronomy, University of Southampton, Highfield SO17 1BJ, UK
e-mail: h.ulbricht@soton.ac.uk

which consist of many atoms, as resonances are manifold and the oscillator strength is distributed across many different state transitions far away from the ideal two-level system situation which we luckily find in some atoms, which gave rise to a revolution in experimental physics. Cold atom experiments now allow for ultra-precise control of various degrees of freedoms—including the centre of mass motion of the atoms and to prepare non-classical sates, including collective ones such as atomic Bose Einstein Condensates (BEC). In a way our programme aims to achieve a similar level of control, but for particles of large mass and different cooling and manipulation techniques have to be developed and used for that. The off-resonant dipole force where α is a measure of the off-resonant detuning of all affected molecular states is however too weak to lead to a large enough effect to trap and manipulate individual molecules by coherent laser light [2, 3] . This situation changes dramatically if one increases the size (volume V) of the particle to trap and therefore its polarizability $\alpha \propto V$. Then dipole force becomes so strong to form a deep optical trap and optical fields can be used for controlling single particle motions again, which gave rise to the development of the new research field levitated optomechnaics [4], based on early pioneering work by Arthur Ashkin (Nobel Prize in Physics in 2018) [5] and already then in close relation to the then soon to be called cold atomic and optical physics. By now the field of levitated large-mass particle systems has seen the implementation of other than optical forces for trapping and manipulation, namely time-varying electrical fields in Paul traps [6] and magnetic traps [7], sometimes including superconductors [8]. All such technical developments give rise to the hope to soon perform experiments with truly macroscopic quantum systems, outperforming existing paradigms of large-mass matterwave interferometry [9]. Macroscopic here entails the involvement of a large-mass particle in a quantum superposition of large spatial separation [10].

There are two pillars of our research programme on testing fundamental physics are with a certain methodological approach. The *first* is the clearly distinctive predictions for the outcome of the same experiment originated from alternative theoretical descriptions. This is our approach for testing the universality of the quantum superposition principle in the context of collapse models [11]. Quantum mechanics and collapse models predict a different outcome of a matterwave interferometry experiment—if the experiment is performed in the right parameter regime. The *second* pillar of our research programme is to first perform a detailed analysis of the new physics to be tested and then to chose the best experiment to perform the test.

Outlook of this chapter. In the following, we will address new avenues to test quantum mechanics in Sect. 2 with the specific emphasis on experiments using levitated mechanical systems. Then we will address experimental tests of the interplay between quantum mechanics and gravity in Sect. 3 including the discussion of the semiclassical Schrödinger-Newton equation, gravitational deocherence of the wavefunction and the gravity of a quantum state. In the final Sect. 4 we will refer to using the Wigner function to simulate the original Stern Gerlach experiment.

2 Testing Quantum Mechanics with Collapse Models

There is an increasing interest in developing experiments aimed at testing collapse models, in particular the Continuous Localization Model (CSL), the natural evolution of the GRW model initially proposed by Ghirardi et al. [11–14]. Current experiments and related bounds on collapse parameters are partially discussed in other contributions in this review. Our aim here is to discuss some of the most promising directions towards future improvements. We will mostly focus on non-interferometric experiments. In Sect. 2.1 we will briefly outline proposals of matter-wave interference with massive nano/microparticles. Finally, in Sect. 2.2 we will discuss mechanical experiments, in particular ongoing experiments with ultracold cantilevers, ongoing and proposed experiments based on levitated nanoparticles and microparticles. We will not consider here two important classes of experiments which are separately discussed by other contributors in this review: matter-wave interference with molecules and space-based experiments. We will end in Sect. 2.3 with some ideas on how precision experiments can be used for testing collapse models. A summary of recent interferometric and non-interferomtric experiments which could set direct bounds on the CSL collapse model are summarized in Fig. 1.

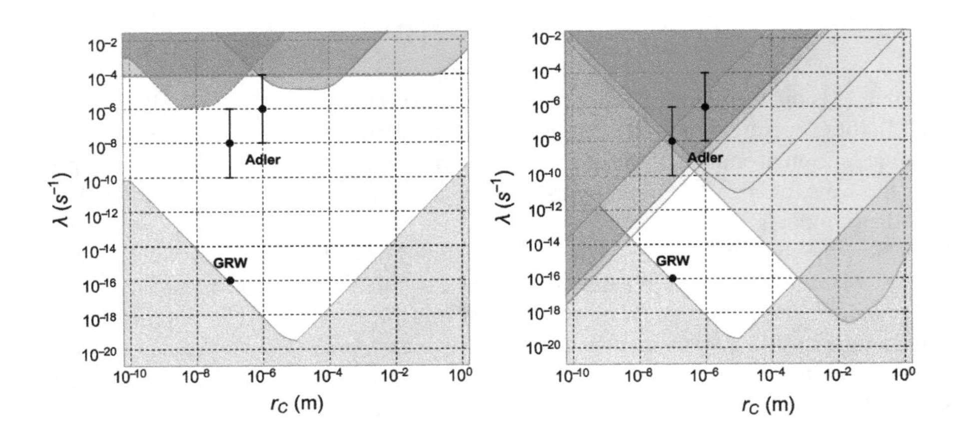

Fig. 1 Exclusion plots for the CSL parameters with respect to the GRW and Adler theoretically proposed values [12, 15]. Left panel—Excluded regions from *interferometric* experiments: molecular interferometry [16] (blue area), atom interferometry [17] (green area) and experiment with entangled diamonds [18] (orange area). Right panel—Excluded regions from *non-interferometric* experiments: LISA Pathfinder [19, 20] (green area), cold atoms [21] (orange area), phonon excitations in crystals [22] (red area), X-ray measurements [23] (blue area) and nanomechanical cantilever [24]. We report with the grey area the region excluded based on theoretical arguments [25]. Figure and caption taken from [26]

2.1 Tests of Quantum Mechanics by Matter-Wave Interferometry

Matterwave interferometry is directly testing the quantum superposition principle. Relevant for mass-scaling collapse models, such as CSL, are matterwave interferometers testing the maximal macroscopic extend in terms of mass, size and time of spatial superpositions of single large-mass particles. Such beautiful, but highly challenging experiments have been pushed by Markus Arndt's group in Vienna to impressive particle masses of 10^4 atomic mass units (amu), which still not significantly challenging CSL. Therefore the motivation remains to push matterwave interferometers to more macroscopic systems. Predicted bounds on collapse models set by large-mass matterwave interferometers are worked out in detail in [27].

As usual in open quantum system dynamics treatments, non-linear stochastic extensions of the Schrödinger equation on the level of the wavefunction [28] correspond to a non-uniquely defined master equation on the level of the density matrix ρ to describe the time evolution of the quantum system, say the spatial superposition across distance $|x - y|$, where the conserving von Neumann term $\partial\rho_t(x, y)/\partial t = -(i/\hbar)[H, \rho]$, is now extended by a Lindblad operator L term:

$$\frac{\partial\rho_t(x, y)}{\partial t} = -\frac{i}{\hbar}[H, \rho_t(x, y)] + L\rho_t(x, y), \tag{1}$$

where H is the Hamilton operator of the quantum system and different realisations of a Lindblad operator are used to describe both standard decoherence (triggered by the immediate environment of the quantum system) [29] as well as spontaneous collapse of the wavefunction triggered by the universal classical noise field as predicted by collapse models.

Now the dynamics of the system is very different with and without the Lindbladian, where with the Lindbladian the unitary evolution breaks down and the system dynamics undergoes a quantum-to-classical transition witnessed by a vanishing of the fringe visibility of the matterwave interferometer. In the state represented by the density matrix the off-diagonal terms vanish as the system evolves according to the open system dynamics, the coherence/superposition of that state is lost. The principal goal of interference experiments with massive particles is then to explore and quantify the relevance of the $(L\rho_t(x, y))$-term—as collapse models predict a break down of the quantum superposition principle for a sufficient macroscopic system. An intrinsic problem is the competition with known and unknown environmental decoherence mechanisms, if a visibility loss is observed. However solutions seem possible.

In order to further increase the macroscopic limits in interference some ambitious proposals have been made utilizing nano- and micro-particles, c.f. Fig. 2. The main challenge is to allow for a long enough free evolution time of the prepared quantum superposition state in order to be sensitive to the collapsing effects. The free evolution—the spatial spreading of the wavefunction $\Psi(r, t)$ with time—according to the time-dependent Schrödinger equation with the potential $V(r) = 0$,

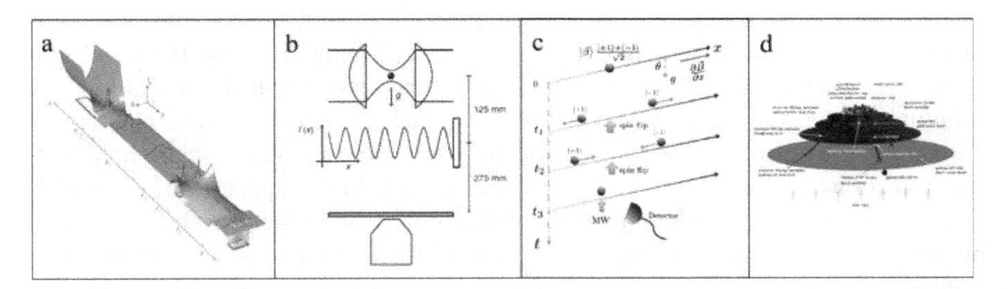

Fig. 2 Illustration of some of the proposed schemes for matterwave interferometry with nano- and micro-particles to test the quantum superposition principle directly, and therefore also collapse models. **a** The cryogenic skatepark for a single superconducting micro-particle (adapted from Ref. [31]); **b** The nanoparticle Talbot interferometer (adapted from Ref. [30]); **c** The Ramsey scheme addressing the electron Spin of a NV-centre diamond coupled to an external magnetic field gradient $(\partial B/\partial x)$ (adapted from Ref. [32]); **d** The adaptation of an interferometer at a free falling satellite platform in space to allow form longer free evolution times (adapted from Ref. [33])

$$\frac{\partial}{\partial t}\Psi(r,t) = -i\frac{\hbar}{2m}\nabla^2\Psi(r,t), \tag{2}$$

describes a diffusive process for probability amplitudes similar to a typical diffusion equation with the imaginary diffusion coefficient $(-i\hbar/2m)$. Therefore the spreading of $\Psi(r, t)$ scales inverse with particle mass m. For instance for a 10^7 amu particle it already takes so long to show the interference pattern in a matterwave experiment that the particle would significantly drop in Earth's gravitational field, in fact it would drop on the order of 100 m. This requires a dramatic change in the way large-mass matterwave interferometry experiments have to be performed beyond the mass of 10^6 amu [30].

Different solutions are thinkable. One could of course envisage building 100 m fountain, but that seems very unfeasible also given that no sufficient particle beam preparation techniques exist (and don't seem to be likely to be developed in the foreseeable future) to enable the launch and detection of particles in the mass range in question over a distance of 100 m. One can consider to levitate the particle by a force field to compensate for the drop in gravity, but here we face a high demand on the fluctuations of that levitating field, which have to be small compared to the amplitudes of the quantum evolution, which is not feasible with current technology. A maybe possible option os to coherently boost/accelerate the evolution of the wave-function spread by a beam-splitter operation. The proposals in Refs. [31, 32] are such solutions, which are still awaiting their technical realisation for large masses. Alternatively and more realistic given technical capability is to allow for long enough free evolution by freely fall the whole interferometer apparatus in a co-moving reference frame with the particle. This is the idea of the MAQRO proposal, a dedicated satellite mission in space to perform large-mass matterwave interference experiments with micro- and nano-particles [33].

Another interesting approach is to consider the use of cold or ultra-cold ensembles of atoms such as cloud in a magneto optical trap (MOT) or an atomic Bose-Einstein Condensate (BEC) as also there we find up to 10^8 atoms of alkali species such as rubidium or caesium. On closer look it turns out that such weakly interacting atomic ensembles are not of immediate use for the purpose to test macroscopic quantum superpositions in the context of collapse model test. For instance testing the CSL model is build on a mass (number of particles N, more precisely the number of nucleons: protons and neutrons in the nuclei of the atoms) amplification which in principle can even go with N^2, if the condition for coherent scattering of the classical collapse noise treated as a wave with correlation length r_c scattered at the particle in the quantum superposition state. The central assumption of this amplification mechanism is that if the CSL noise is collapsing the wavefunction of only one of the constituent nucleons, then the total wavefunction of the whole composite object collapses. While in the case of a nanoparticle consisting of many atoms (and therefore nucleons), it is not the case for an weakly interacting atomic ensemble. If one atom is collapsing then the total atomic wavefunction remains intact and the one atom is lost from the ensemble.

This may change if the atoms in the cold or ultra-cold ensemble can be made stronger interacting, without running into the complications of chemistry which may forbid condensation of the atomic—then molecular—cloud at all. However there is hope that quantum optical state preparation techniques applied after a BEC has been formed such as collective NOON or squeezed states enable N and even N^2 scaling in the fashion fit for testing wavefunction collapse.

Interestingly, this might be different if the physical mechanism responsible for the collapse of the wavefunction, which remains highly speculative at present, is in any way related to gravity [34], then there might be hope that atomic ensembles even in the weakly interacting case can be used to test CSL-type models. The condition to fulfil is that the atomic ensemble is interacting gravitationally strong enough so that it acts collectively under collapse, even if just a single constituent atom (nucleon) is affected by the collapsing effect.

2.2 *Non-interferometric Mechanical Tests of Quantum Mechanics*

This class of experiments has emerged in recent years as one of the most powerful and effective ways to test collapse models. The underlying idea [35, 36] is that a mechanism which continuously localizes the wavefunction of a mechanical system, which can be either a free mass or a mechanical resonator, must be accompanied by a random force noise acting on its center-of-mass. This leads in turn to a random diffusion which can be possibly detected by ultrasensitive mechanical experiments.

In a real mechanical system such diffusion will be masked by standard thermal diffusion arising from the coupling to the environment, i.e. from the same effects which lead to decoherence in quantum interference experiments [37]. In practice there will be additional non-thermal effects, due to external non-equilibrium vibrational noise (seismic/acoustic/gravity gradient). Moreover, one has to ensure that the back-action from the measuring device is negligible.

Under the assumption that thermal noise is the only significant effect, the (one-sided) power spectral density of the force noise acting on the mechanical system is given by:

$$S_{ff} = \frac{4k_B T m\omega}{Q} + 2\hbar^2\eta. \tag{3}$$

where k_B is the Boltzmann constant, T is the temperature, m is the mass, ω the angular frequency, Q is the mechanical quality factor.

η is a diffusion constant associated to spontaneous localization, and can be calculated explicitly for the most known models. For CSL, it is given by the following expression

$$\eta = \frac{2\lambda}{m_0^2} \int\int \mathrm{d}^3\mathbf{r}\,\mathrm{d}^3\mathbf{r}' \exp\left(-\frac{|\mathbf{r}-\mathbf{r}'|^2}{4r_C^2}\right) \frac{\partial\varrho(\mathbf{r})}{\partial z}\frac{\partial\varrho(\mathbf{r}')}{\partial z'} \tag{4}$$

$$= \frac{(4\pi)^{\frac{3}{2}}\,\lambda\, r_C^3}{m_0^2} \int \frac{\mathrm{d}^3\mathbf{k}}{(2\pi)^3} k_z^2\, e^{-\mathbf{k}^2 r_C^2}\, |\tilde{\varrho}(\mathbf{k})|^2 \tag{5}$$

with $\mathbf{k} = (k_x, k_y, k_z)$, $\tilde{\varrho}(\mathbf{k}) = \int \mathrm{d}^3\mathbf{x}\, e^{i\mathbf{k}\cdot\mathbf{r}}\,\varrho(\mathbf{r})$ and $\varrho(\mathbf{r})$ the mass density distribution of the system. In the expressions above m_0 is the nucleon mass and r_C and λ are the free parameters of CSL. The typical values proposed in CSL literature are $r_C = 10^{-7}$ m and 10^{-6} m, while for λ a wide range of possible values has been proposed, which spans from the GRW value $\lambda \approx 10^{-16}$ Hz [12, 13] to the Adler value $\lambda \approx 10^{-8\pm2}$ Hz at $r_C = 10^{-7}$ m [15]. The possibility for such non-interferometric tests, which aim to directly test the non-thermal noise predicted by collapse models [24, 38–41].

An experiment looking for CSL-induced noise has to be designed in order to maximize the 'noise to noise' ratio between the CSL term and the thermal noise. In practice this means lowest possible temperature T, lowest possible damping time, or linewidth, $1/\tau = \omega/Q$, and highest possible η/m ratio. The first two conditions express the requirement of lowest possible power exchange with the thermal bath, the third condition is inherently related to the details of the specific model.

For CSL we can distinguish two relevant limits. When the characteristic size L of the system is small, $L \ll r_C$, then the CSL field cannot resolve the internal structure of the system, and one finds $\eta/m \propto m$. When the characteristic length of the system in the direction of motion L is large, $L \gg r_C$, then $\eta/m \propto \rho/L$, where ρ is the mass density [24, 38, 40]. The expressions in the two limits imply that, for a well defined characteristic length r_C, the optimal system is a plate or disk with thickness $L \sim r_C$ and the largest possible density ρ.

Among other models proposed in literature, we mention the gravitational Diosi-Penrose (DP) model, which leads to localization and diffusion similarly to CSL. The diffusion constant η_{DP} is given by [40]:

$$\eta_{DP} = \frac{G\rho m}{6\sqrt{\pi}\hbar}\left(\frac{a}{r_{DP}}\right)^3, \tag{6}$$

where a is the lattice constant and G is the gravitational constant, so that he ratio η_{DP}/m depends only on the mass density. Unlike CSL, there is no explicit dependence on the shape or size of the mechanical system.

2.2.1 Levitated Mechanical Systems

One of the most promising approaches towards a significant leap forward in the achievable sensitivity to spontaneous collapse effects is by levitation of nanoparticles or microparticles. The main benefits of levitation are the absence of clamping mechanical losses and wider tunability of mechanical parameters. In addition, several degrees of freedom can be exploited, either translational or rotational [41, 42]. This comes at the price of higher complexity, poor dynamic range and large nonlinearities, which usually require active feedback stabilization over multiple degrees of freedom. However, levitated systems hold the promise of much better isolation from the environment, therefore higher quality factor. One relevant example, in the macroscopic domain, is the space mission LISA Pathfinder, which is based on an electrostatically levitated test mass, currently setting the strongest bound on collapse models over a wide parameter range [43].

Several levitation methods for micro/nanoparticles are currently being investigated. The most developed is optical levitation using force gradients induced by laser fields, the so called optical tweezer approach [5]. While this is a very effective and flexible approach to trap nanoparticles, in this context it is inherently limited by two factors: the relatively high trap frequency, in the order of 100 kHz, and the high internal temperature of the particles, induced by laser power absorption, which leads ultimately to strong thermal decoherence. Alternative approaches have to be found, featuring lower trap frequency and low or possible null power dissipated in the levitated particle. The two possible classes of techniques are electrical levitation and magnetic levitation.

Electrical levitation has been deeply developed in the context of ion traps. The standard tool is the Paul trap, which allows to trap an ion, or equivalently a charged nanoparticle, using a combination of ac and dc bias electric fields applied through a set of electrodes [44]. The power dissipation is much lower than in the optical case, and the technology is relatively well-established. However, the detection of a nanoparticle in a Paul trap still poses some technological challenge (Fig. 3).

This issue has been extensively investigated in a recent paper [45], specifically considering a nanoparticle in a cryogenic Paul trap in the context of collapse model testing. Three detection schemes have been considered: an optical cavity, an optical tweezer, and a all-electric readout based on SQUID. It was found that to detect the nanoparticle motion with good sensitivity, optical detection has to be employed. Unfortunately, optical detection is not easily integrated in a cryogenic environment, and leads to a nonnegligible internal heating and excess force noise. On the other

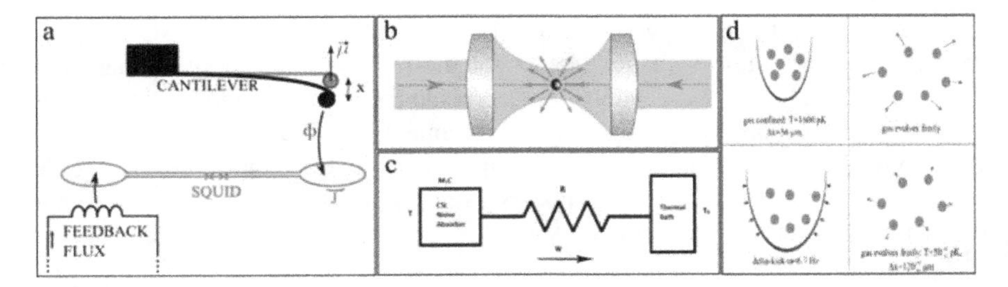

Fig. 3 Simplified sketch of some of the noninterferometric methods to test collapse models discussed in this contribution. **a** Measuring the mechanical noise induced by CSL using an ultracold cantilever detected by a SQUID (adapted from Ref. [24]); **b** Measuring the mechanical noise induced by CSL using a levitated nanoparticle detected optically (adapted from Ref. [45]); **c** Measuring the heating induced by CSL in a solid matter object cooled to very low temperature (adapted from Ref. [46]); **d** Measuring the increase of kinetic energy induced by CSL in a ultracold atoms cloud (adapted from Ref. [47])

hand, an all-electrical readout would potentially allow for a better ultimate test of collapse models, but at the price of a very poor detection sensitivity, which could make the experiment hardly feasible. The authors argue that a Paul-trapped nanoparticle, with an oscillating frequency of 1 kHz, cooled in a cryostat at 300 mK with an optical readout may be able to probe the CSL collapse rate down to 10^{-12} Hz at $r_C = 10^{-7}$ m. A SQUID-based readout , if viable, could theoretically allow to reach 10^{-14} Hz.

A recent experiment employing a nanoparticle in a Paul trap with very low secular frequencies at ~100 Hz and low pressure has demonstrated ultranarrow linewidth $\gamma/2\pi = 82$ μHz [48]. This result has been used to set new bounds on the dissipative extension of CSL. This experiment may be able to probe the current limits on the CSL model in the near future, once it will be performed at cryogenic temperature and the main sources of excess noise, in particular bias voltage noise, will be removed.

Magnetic levitation, while less developed, has the crucial advantage of being completely passive. Furthermore the trap frequencies can be quite low, in the Hz range. Three possible schemes can be devised: levitation of a diamagnetic insulating nanoparticle with strong external field gradients [49, 50], levitation of a superconducting particle using external currents [8, 51–53], and levitation of a ferromagnetic particle above a superconductor [54].

The first approach has been recently considered in the context of collapse models [50]. The experiment was based on a polyethylene glycol microparticle levitated in the static field generated by neodymium magnets and optical detection. The experiment has been able to set an upper bound on the CSL collapse rate $\lambda < 10^{-6.2}$ Hz at $r_C = 10^{-7}$ m, despite being performed at room temperature. A cryogenic version of this experiment should be able to approach the current experimental limits on CSL.

The second and third approach based on levitating superconducting particles are currently investigated by a handful of groups [8, 51–54], but no experiment has so far reached the experimental requirements needed to probe collapse models. However, a significant progress has been recently achieved: a ferromagnetic microparticle levitated above a type I superconductor (lead) and detected using a SQUID, has demonstrated mechanical quality factors for the rotational and translational rigid body mechanical modes exceeding 10^7, corresponding to a ringdown time larger than 10^4 seconds [53]. The noise is this experiment is still dominated by external vibrations. However, as the levitation is completely passive and therefore compatible with cryogenic temperatures, this appears as an excellent candidate towards near future improved tests of collapse models.

2.3 Concluding Remarks on Testing Quantum Mechanics in the Context of Collapse Models

We have discussed avenues for non-interferometric and interferometric tests of the linear superposition principle of quantum mechanics in direct comparison to predictions from collapse models which break the linear/unitary evolution of the wavefunction. As matters stand both non-interferometric and interferometric set already bounds on the CSL collapse model, while those from non-interferometric tests are stronger by orders of magnitude. The simple reason lyes in the immense difficulty to experimentally generate macroscopic superposition states, however a number of proposals have been made and experimentalists are set to approach the challenge.

We want to close by mentioning that there are possible other experimental platforms which could set experimental bunds on collapse models and it would be of interest to study those in detail. Collapse models predict a universal classical noise field to fill the Universe and in principle couple to any physical system. In the simplest approach the experimental test particle can be regarded as a two-level system, as typically described in quantum optics. Then the collapse noise perturbs the two-level system and emissive broadening and spectral shifts can be expected, unfortunately out of experimental reach at the moment [55]. The minuscule collapse effect on a single particle (nucleon) needs some sort of amplification mechanism which usually comes with an increase of the number of constituent particles. However, ultra-high precision experiments have improved a lot in recent years. For instance much improved ultra-stable Penning ion traps are used to measure the mass of single nuclear particles, such as the electron, proton, and neutron, with an ultra-high precision to test quantum electrodynamics predictions [56]. In principle also here the effect of collapse models should become apparent. Any theoretical predictions are difficult as relativistic versions of collapse models still represent a serious formal challenge. Other high potentials for testing collapse are ever more precise spectroscopies of simple atomic species with analytic solutions such as transitions in hydrogen [57] and needless to say atomic clocks [58].

As tests move on to set stronger and stronger bounds, we have to remain open to actually find something new. It is so easy to disregard tiny observed effects as unknown technical noise. In the case of direct testing collapse noise it is a formidable theoretical challenge to think about possible physics responsible for collapse, satisfying the constrains given by the structure of the collapse equation: the noise has to be classical and stochastic. Such concrete physics models will predict a clear frequency fingerprint, should we ever observe the collapse noise field.

3 Testing the Interplay Between Quantum Mechanics and Gravity

Here we will be concerned with table-top experiments in the non-relativistic regime as these experiments may provide a new access to shine light on the quantum—gravity interplay. Therefore the main emphasis is to explore possible routes to enter the new parameter regime, where both quantum mechanics and gravity are significant, see Fig. 4). This means the mass of the object has to be large enough to show gravity effects while also not being too large to still allow for the preparation of non-classical features of the behaviour of that massive object. That regime where both physical effects, the quantum and the gravity, could be expected to be relevant is at around the Planck mass, which is derived from the right mixture of fundamental constants ($\hbar$ Planck's constant, c speed of light, G gravitational constant) $m_{pl} = \sqrt{\hbar c/G} = 2.176470(51) \times 10^{-8}$ kg (the official CODATA, NIST) or below. No quantum experiment has been performed in that mass range.

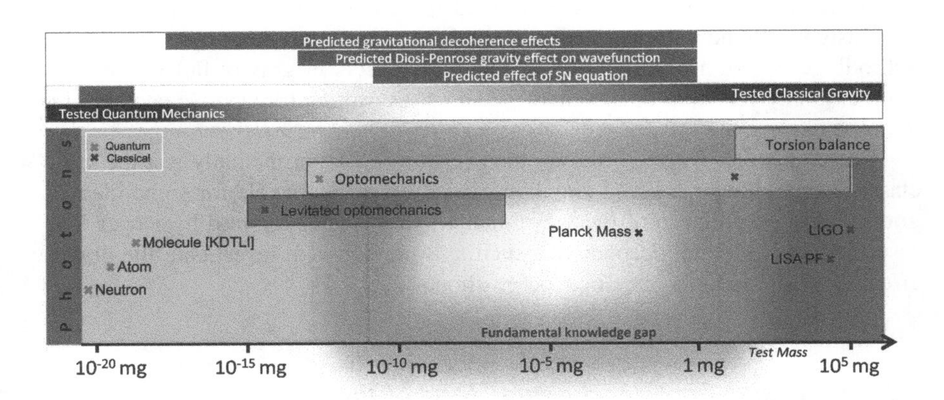

Fig. 4 Exploration map of mass: Mass range of the test mass as explored by experiments. Experiments to detect gravity have been done in the classical domain, *right hand side of picture*, with comparable large masses. Quantum experiments are routinely performed by using objects of much smaller masses so that gravity effects do not become visible or relevant. Neutron and atom matter-wave interferometers are different as the test mass there is very small [the mass of a single neutron or atom], but in a spatial superposition state. The desired mass range for—at least some of—the experiments summarized in this review article is at the overlap between sufficiently large mass to see significant effects of gravity of the particle itself, while the particle can be maintained in a non-classical state. The domain where massive particles can be prepared in such non-classical states is *on the left hand side* of the picture

When we refer to quantum mechanical behaviour of massive systems, we mean the centre of mass motion of such a system, which may consist of many atoms. Surely, there are many other [we call those internal] degrees of freedom of the same system such as electronic states or vibrations and rotations which are described as relative motions of the atoms forming the large object, but here we are not concerned with those. When we talk about superpositions, we mean spatial superpositions, in the sense of a the centre of mass of a single particle, which can be elementary or composite, being *here* and *there* at a given time, the Schrödinger cat state. The most massive complex quantum system, which has been experimentally put in such a superposition state, are complex organic molecules of a mass on the order of $m_{max} = 10^{-22}$ kg [9].

Typically for gravity experiments there are two masses involved, the source mass which generates a gravitational field, potential or curvature of space-time (the source mass has usually a big mass) and the test mass which is probing the gravity effect generated by the source mass. Torsion balances are the classic device for typical gravity experiments. We think there are two regimes interesting for experimental investigation: (1) the regime where a quantum system is the test mass and interacts with a large external source mass. This is the regime where neutron and atom interferometry are already very successful and provide tools for precise measurements of gravity effects. (2) the regime where the quantum system itself carries sufficient mass to be the source mass and to allow for related quantum gravity effects to become experimentally accessible. So far there has been no convincing experiment in the second regime. Any experiment performed in that second regime will ultimately give insight into the interplay between gravity and quantum mechanics. Test of the Schrödinger-Newton equation and of quantum effects in gravity fall in the latter regime. It may very well be that there are surprises waiting for us if we become able to probe that regime by experiments.

In the following we shall discuss the prospects to experimentally test the semi-classical Schrödinger-Newton equation, which plays also a role for some ideas of gravity induced collapse of the wavefunction such as put forward by Roger Penrose [59], gravitational decoherence such as some ideas to investigate the gravity effects within a spatial quantum superposition state.

3.1 Proposals for Experimental Tests of the Schrödinger-Newton equation

What is the gravitational field of a quantum system in a spatial superposition state? The seemingly most obvious approach, the perturbative quantization of the gravitational field in analogy to electromagnetism, makes it alluring to reply that the space-time of such a state must also be in a superposition. The non-renormalizability of said theory, however, has also inspired the hypothesis that a quantization of the gravitational field might not be necessary after all [60, 61]. Rosenfeld already expressed the thought that the question whether or not the gravitational field must be quantized can only be answered by experiment: *There is no denying that, considering the universality of the quantum of action, it is very tempting to regard any classical theory as a limiting case to some quantal theory. In the absence of empirical evidence, however, this temptation should be resisted. The case for quantizing gravitation, in particular, far from being straightforward, appears very dubious on closer examination.* [60]

Adopting this point of view, an alternative approach to couple quantum matter to a classical space-time is provided by a fundamentally semi-classical theory that is by replacing the source term in Einstein's field equations for the curvature of classical space-time, energy-momentum, by the *expectation value* of the corresponding quantum operator [60, 62]:

$$R_{\mu\nu} + \frac{1}{2} g_{\mu\nu} R = \frac{8\pi G}{c^4} \langle \Psi | \hat{T}_{\mu\nu} | \Psi \rangle. \tag{7}$$

Of course, such presumption is not without complications. For instance, in conjunction with a no-collapse interpretation of quantum mechanics it would be in blatant contradiction to everyday experience [63]. Moreover, the nonlinearity that the back-reaction of quantum matter with classical space-time unavoidably induces cannot straightforwardly be reconciled with quantum nonlocality in a causality preserving manner [64, 65]. Be that as it may, there is no consensus about the conclusiveness of these arguments [66–68]. The enduring quest for a theory uniting the principles of quantum mechanics and general relativity gives desirability to having access to hypotheses which could be put to an experimental test in the near future.

In the non-relativistic limit, the assumption of fundamentally semi-classical gravity yields a non-linear, nonlocal modification of the Schrödinger equation, commonly referred to as the Schrödinger–Newton equation [34, 69, 70]. After a suitable approximation [70], for the center of mass of a complex quantum system of mass M in an external potential V_{ext} it reads:

$$\mathrm{i}\hbar \frac{\partial}{\partial t} \psi(t, \mathbf{r}) = \left(\frac{\hbar^2}{2M} \nabla^2 + V_{\text{ext}} + V_g[\psi] \right) \psi(t, \mathbf{r}) \tag{8a}$$

$$V_g[\psi](t, \mathbf{r}) = -G \int \mathrm{d}^3 r' \, |\psi(t, \mathbf{r}')|^2 \, I_{\rho_c}(\mathbf{r} - \mathbf{r}'). \tag{8b}$$

The self-gravitational potential V_g depends on the wavefunction, and hence renders the equation nonlinear. The function I_{ρ_c}, which models the mass distribution of the considered system, will be defined below.

The Schrödinger–Newton equation has primarily been discussed in the context of gravitationally induced quantum state reduction [71, 72]. Its relevance for a possible experimental test of the necessity to quantize the gravitational field was pointed out by Carlip [61]. First ideas how to test such kind of nonlinear, self-gravitational effects focused on the spreading of a free wavefunction in matter-wave interferometry experiments [9, 37, 61, 69, 70]. Recently, other experimental test have been proposed including one based on the internal dynamics of a squeezed coherent ground state of a micron-sized silicon particle in a harmonic potential. We will now discuss further ideas for testing the Schrödinger-Newton equation.

3.1.1 Proposed Direct Tests of Schrödinger-Newton Equation: Wavefunction Expansion

The direct test of the Schrödinger-Newton (SN) equation is by studying the free expansion of the wavefunction of sufficiently massive objects. Then a contraction of the wave function according to the SN self-gravity effect should have a consequence on that expansion, competing with its natural Schrödinger's dynamics spread. Clearly, because to the weakness of gravitation interaction, the mass has to be sufficiently large while the object has to remain in a state which can be described by a centre of mass quantum wavefunction, meaning the spatial extent of the wavefunction should be detectable for the full duration of the evolution. See Fig. 5 for the mass-time parameter space required to observe the predicted SN effect directly, which has been studied extensively. While analytic solutions of the SN equation are difficult and even numerical simulations are non-trivial.

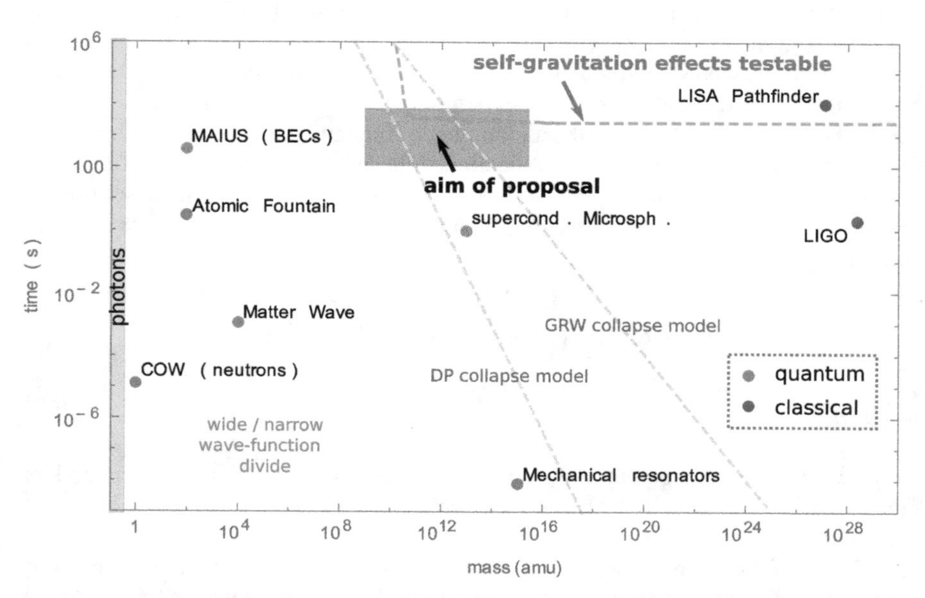

Fig. 5 Direct Test of Schrödinger-Newton (SN) wavefunction evolution: The mass-time plot to illustrate the parameter range which needs to be reached for direct SN wavefunction evolution experiments. This clearly needs to be done without external gravity and other forces/interactions and therefore an experiment in space appears a likely option. The red area shows the parameter range for a proposed space mission to test the SN effect

One possible experimental scenario would be a molecule interferometry experiment [9]. While such matterwave experiments probe spatial superposition states of large molecules—the SN contraction effect could also be observed for a free expansion of a singular wave function originated from a point in space. The key is that the mass of the evolving quantum object has to be comparable large, much larger than the mass achieved in present molecule interferometry experiments. Cold atoms and even BEC of atoms, which benefit from the multitude of coherent manipulation, control and cooling schemes do not seem to have large enough mass in order to show the SN expansion/contraction effect. Clearly one needs a high mass at the same time as access to the coherent quantum evolution of the objects wavefunction. The high mass and the long expansion times to be studied challenge the experimental realisation.

Therefore, should direct tests of the SN equation be done in space? Yes, at this point there seems to be no other way to allow the wavefunction expansion for long enough, typically some hundred seconds, see Fig. 5. Proposals to levitate massive particles (optically or magnetically) and therefore to compensate for the drop in Earth's gravity have not been realised and are more problematic for SN test. The levitated tests rely on proposed techniques to accelerate the wavefunction expansion artificially by optical or magnetic field gradients. That acceleration would have the potential to wash out completely the fragile SN effect.

3.1.2 Proposed Indirect Tests of SN Equation

Indirect SN effects have been predicted for optomechanics systems which are comparably massive and on the verge to be quantum, see Fig. 4. Such effects are very small, can be overwhelmed by noise effects in the experiments, but can be done on the table-top. Therefore these tests represent a serious experimental challenge, while proposed to be possible with available technology. Two optomechanics experimental cases and the study of the SN dynamics in non-linear optics analogs are mentioned:

A. SN rotation of squeezed states The mechanical motion of an optomechanical system, clamped or levitated, is squeezed. Quantum squeezing of clamped optomechanics has been realised experimentally already, while a classical analog has been demonstrated for a levitated system. An optical homodyne detection of both field quadratures of the mechanical state is plotted and shows the cigar-shaped state, see in Fig. 6 left. The SN equation predicts an extra rotation of the squeezed phase-space distribution [73].

B. SN energy shifts of mechanical harmonic oscillator A further theoretical study [74] predicts SN related shifts of the Eigenenergy levels of the quantum harmonic oscillator describing the optomechanical system, see for an illustration of the multiple energy shift effects the Fig. 6 right. There different effects for the so-called wide and narrow wavefunction regimes are predicted for the situations that the spatial extent of the centre of mass motion wavefunction is larger (wide wavefunction regime) or smaller (narrow wavefunction regime) than the physical size of the massive object. A detailed experimental scenario has been worked out and awaits its realisation in an actual laboratory.

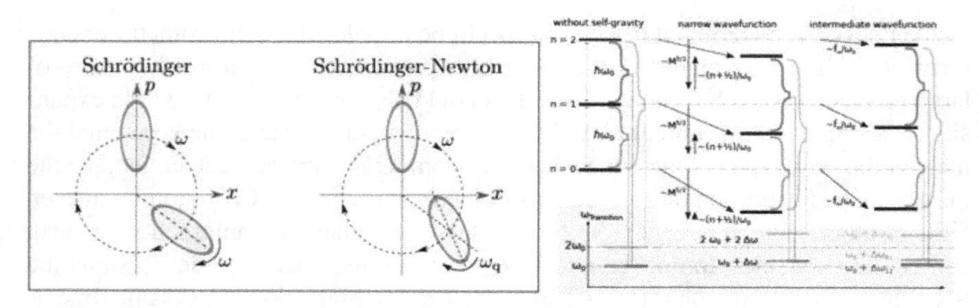

Fig. 6 Indirect Tests of the Schrödinger-Newton equation: *Left Panel*: Phase space plot of mechanical squeezed state with extra rotation of state distribution according to the SN effect. Left side: according to standard quantum mechanics, both the vector $(\langle x \rangle, \langle p \rangle)$ and the uncertainty ellipse of a Gaussian state for the centre of mass (CM) of a macroscopic object rotate clockwise in phase space, at the same frequency $\omega = \omega_{CM}$. Right side: according to the CM Schrödinger-Newton equation, $(\langle x \rangle, \langle p \rangle)$ still rotates at ω_{CM}, but the uncertainty ellipse rotates at $\omega_q = (\omega_{CM}^2 + \omega_{SN}^2)^2$. Picture taken from [73]. *Right Panel:* Schematic overview of the effect of the Schrödinger-Newton equation on the spectrum. The top part shows the first three energy eigenvalues and their shift due to the first order perturbative expansion of the Schrödinger-Newton potential. The bottom part shows the resulting spectrum of transition frequencies. In the narrow wavefunction regime (middle part), all energy levels are shifted down by an n-independent value minus an n-proportional contribution that scales with the inverse trap frequency. In the intermediate regime, where the wavefunction width becomes comparable to the localization length scale of the nuclei, this n-proportionality does no longer hold, leading to a removal of the degeneracy in the spectrum. Picture and caption taken from [74]

C. Non-linear optics simulation of the SN equation Specific delocalised non-linearities in optical systems, typically just a piece of glass with a large refractive index, show a very similar type of dynamics for the propagation of light though that system if compared to SN dynamics. The analog holds at least in (1+1) space-time dimensions. The analog provides an interesting option to study the dynamics of the SN equation in a parameter regime complementary to numeric simulations. Some experiments have been already performed [75, 76] to study cosmological settings of the SN equation such as exotic Boson stars. The main question remains, what can we ultimately learn from optics analog experiments. Do we really learn about gravity? No, but we learn about the formal analog dynamics which is hard to calculate or simulate otherwise.

3.2 Gravitational Decoherence Effects

Tests of gravitational decoherence are based on the the straight-forward approach to generate a spatial superposition state (or any other non-classical state) of a massive particle and test if such a state decoheres according to (classical or quantum) gravity. Clearly, the experimental challenge is the preparation of such a state of sufficient mass. Typical experiments involve matterwave interferometers and quantum optomechanics. While the larges mass is given again by molecule interferometry—some of the effects (such as time dilation) are more promising to be tested in smaller mass systems such as cold atom interferometry, as those can be prepared in larger size superposition states to pick up a larger dephasing or decoherence effect. While on first sight it appears that only massive systems can be used for the test, we know that GR effects also exist for photons [77].

A. Gravitational decoherence affecting superpositions One of the proposed effects is by GR time dilation [78, 79], which is picked up as a dephasing effect for a matterwave interferometer for the propagation of the wavefunction along the two different arms—ultimately resulting in a reduction of the visibility of the interference pattern. The effect has been predicted to scale with the number of all internal degrees of freedom, which are involved in the energy-momentum tensor on the right hand side of Einstein's equations and therefore to affect the spacetime curvature and therefore gravity.

Atom interferometry tests, profiting from the high control on the centre of mass motion of cold atoms, e.g. in 10 m fountain and with sensitively on the verge of 10^{-19}, of the time delation effect appear most promising at the moment, while the theoretical details of the effect are still debated. As a universal decoherence effect to explain the evident macroscopic quantum to classical transition, it is clear that that time dilation decoherence should it exist is weaker by many order of magnitude than know environmental effects such as decoherence due to collisions by an even very diluted background gas [80], which leaves the usefulness of the GR effect in question.

To be more precise, each (internal) degree of freedom of the particle is regarded as a clock running at a typical frequency, but depending via GR time dilation on the local gravitational environment. Then each single clock if separated between the two different paths of an interferometer will be sensitive to the relative duration of time and therefore dephase. This experiment has been realised as a proof of principle experiment with atomic chips [81], where the much larger spatial separation in other atomic interferometers [17] will help to improve the sensitivity to observe the predicted effect to test whether GR time dilation can be regarded as a universal source of decoherence to explain the macroscopic quantum to classical transition of physical systems, ultimately to explain the existence of the classical world.

B. Gravitational effect in dynamical reduction models Dynamical reduction or collapse models have been formulated to explain the quantum to classical transition on a fundamental level and in complement to decoherence models [14]. While the physics reason for the collapse to occur is explained by the existence of a universal classical and random noise field, the physics origin of that field is still debated. Gravity to be a candidate for the collapse field has to fulfil that two conditions of being classical and random. While the classicality is more straight forward, the implementation of a generic stochastic version of gravity represents a challenge. Some attempts have been undertaken and can also been seen as a stochastic modification of the Schrödinger-Newton equation, which was discussed in Sect. 3.1 [34, 82–84]. Tests of such gravity collapse models follow the same logic as tests of collapse models and in general a set of parameters has to be fulfilled. For more details related to

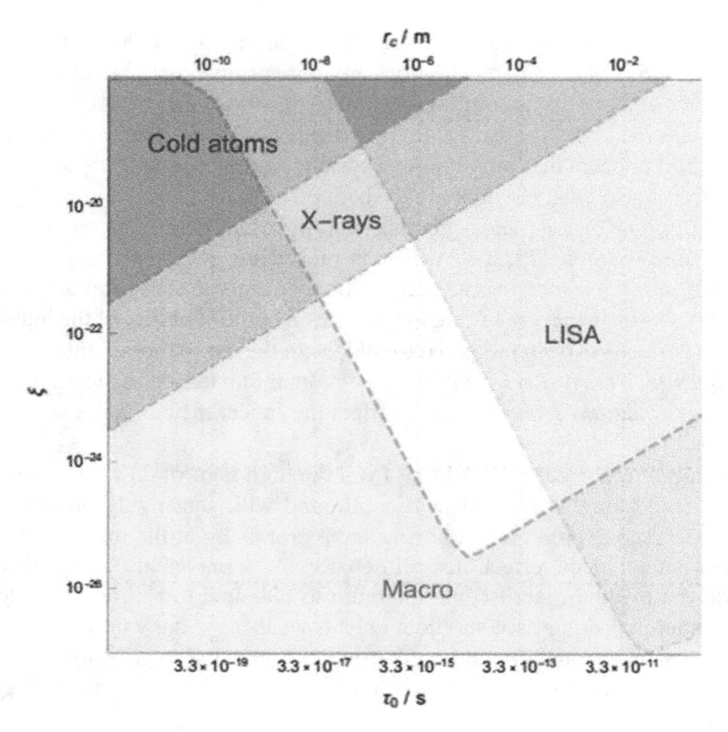

Fig. 7 Parameter map for gravity induced collapse models: (ξ, r_C) or equivalently (ξ, τ_0) parameter diagram of the gravity-induced collapse model. The white area is the allowed region. The blue shaded region (X-rays) is excluded by data analysis of X-rays measurements. The orange shaded region (LISA) is excluded from data analysis of LISA Pathfinder. The green shaded region (Macro) is an estimate of the region excluded by the requirement that the collapse is strong enough to localize macroscopic objects. Note that X-ray measurements sample the high frequency region of the spectrum (10^{18} Hz) and would disappear if the noise correlator has a cutoff below such frequencies, which is plausible. In such a case, the stronger upper bound on the left part of the plane is given by data analysis with cold atom experiments (Cold atoms) [35]. Picture and caption has been taken from [84]

experimental test we refer to [84], where the Fig. 7) has been taken from. While the bounds on gravity collapse models in Fig. 7 relate to experiments already done, future experiments proposed to close the raining gap in the parameter plot involve those to generate large and massive quantum superpositions [30–32, 85]. Such experiments are currently under development in the laboratories.

C. Gravity induced collapse of the wavefunction Other ideas which are less related to the formalism of collapse models, but do explain the collapse of the wavefunction according to gravity are those independently by Diosí [86] and Penrose [87]. The best way to test those models is by large mass matterwave interferometry, where the mass has to be beyond the presently reached limit of molecule interferometry by many orders of magnitude. This means to test such models requires to preparation of large masses in non-classical states and optomechanical or magnetomechanical systems look most promising for the test [88–90]. Proposed experiments along those lines involve [31, 91].

D. Competing effects for matterwave interferometry In order to be able to see such gravity effects and how they collapse or decohere the wavefunction in matter-wave based experiments all competing environmental decoherence processes have to be suppressed, which is the major experimental challenge in order to perform the experiments. Dominating decoherence effects are due to collisions with background gas, collisional decoherence [92] and the effects because of exchange of thermal radiation between the quantum system and the environment [30, 85]. Magnetic levitation of superconducting microparticles by definition avoids all effects related to internal temperature radiation as the experiment is cryogenic and on top of that all noises related to lasers are removed as well [31] which represents a huge advantage compared to optomechanics test. Further vibrations set serious constraints to all mechanics based test of wavefunction collapse and gravity.

E. The case for space Ultimately a test of gravity decoherence and gravity induced collapse of the wavefunction would benefit from large masses of the particles in superposition states as well as long lifetimes of those superposition states in order to observe the extremely weak effects. The space proposal on macroscopic quantum resonators (MAQRO) [33] would be able to fulfil such all those conditions. A community has started to work towards such a test in space and to propose a related mission.

3.3 The Gravity of a Quantum State—Revisited

What gravitational field is generated by a massive quantum system in a spatial superposition? Despite decades of intensive theoretical and experimental research, we still do not know the answer. On the experimental side, the difficulty lies in the fact that gravity is weak and requires large masses to be detectable. However, it becomes increasingly difficult to generate spatial quantum superpositions for increasingly large masses, in light of the stronger environmental effects on such systems. Clearly, a delicate balance between the need for strong gravitational effects and weak decoherence should be found. We show that such a trade off could be achieved in an optomechanics scenario that allows to witness whether the gravitational field generated by a quantum system in a spatial superposition is in a coherent superposition or not. We estimate the magnitude of the effect and show that it offers perspectives for observability.

Quantum field theory is one of the most successful theories ever formulated. All matter fields, together with the electromagnetic and nuclear forces, have been successfully embedded in the quantum framework. They form the standard model of elementary particles, which not only has been confirmed in all advanced accelerator facilities, but has also become an essential ingredient for the description of the universe and its evolution.

In light of this, it is natural to seek a quantum formulation of gravity as well. Yet, the straightforward procedure for promoting the classical field as described by general relativity, into a quantum field, does not work. Several strategies have been put forward, which turned into very sophisticated theories of gravity, the most advanced being string theory and loop quantum gravity. Yet, none of them has reached the goal of providing a fully consistent quantum theory of gravity.

At this point, one might wonder whether the very idea of quantizing gravity is correct [59–64, 66, 66–68, 93, 94]. At the end of the day, according to general relativity, gravity is rather different from all other forces. Actually, it is not a force at all, but a manifestation of the curvature of spacetime, and there is no obvious reason why the standard approach to the quantization of fields should work for spacetime as well. A future unified theory of quantum and gravitational phenomena might require a radical revision not only of our notions of space and time, but also of (quantum) matter. This scenario is growing in likeliness [95–97].

From the experimental point of view, it has now been ascertained that quantum matter (i.e. matter in a genuine quantum state, such as a coherent superposition state) couples to the Earth's gravity in the most obvious way. This has been confirmed in neutron, atom interferometers and used for velocity selection in molecular interferometry. However, in all cases, the gravitational field is classical, i.e. it is generated by a distribution of matter (the Earth) in a fully classical state. Therefore, the plethora of successful experiments mentioned above does not provide hints, unfortunately, on whether gravity is quantum or not.

In a recent paper [98], we discuss an approach where a quantum system is forced in the superposition of two different positions in space, and its gravitational field is explored by a probe (Fig. 8). Using the exquisite potential for transduction offered by optomechanics, we can in principle witness whether the gravitational field is the

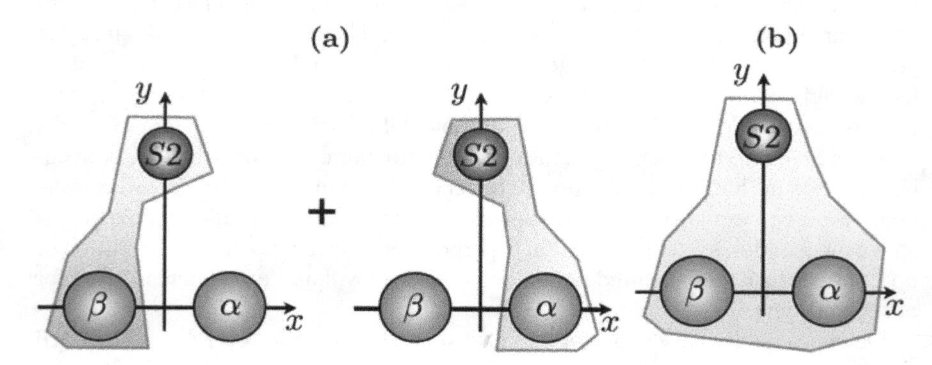

Fig. 8 Schematic representation of the two-body setup. S1 is prepared in a spatial superposition along the x direction (red balls). S2 is initially prepared in a localized wavepacket (blue ball), and it probes the gravitational field generated by S1. **a** The gravitational field acting on S2 is a linear combination of gravitational fields produced by S1 being in a superposed state. **b** The semi-classical treatment of gravity, where the gravitational field acting on S2 is that produced by a total mass m_1 with density $\frac{1}{2}\left(|\alpha(r)|^2 + |\beta(r)|^2\right)$

superposition of the two gravitational fields associated to the two different states of the system, or not. The first case amounts to a quantum behavior of gravity, the second to a classical-like one. We have illustrated the dynamics of an optomechanical system probing the gravitational field of a massive quantum system in a spatial superposition. Two different dynamics are found whether gravity is treated quantum mechanically or classically. Here, we propose two distinct methods to infer which of the two dynamics rules the motion of the quantum probe, thus discerning the intrinsic *nature* of the gravitational field. Such methods will be then eventually able to falsify one of the two treatments of gravity. A similar proposal has been made for angular superpositions [99].

The considered setup is formed of two systems interacting gravitationally. All non-gravitational interactions are considered, for all practical purposes, negligible. The first system (S1) has a mass m_1, and it is initially prepared in a spatial superposition along the x direction. Its wave-function is $\psi(r_1) = \frac{1}{\sqrt{2}}(\alpha(r_1) + \beta(r_1))$, where $\alpha(r_1)$ and $\beta(r_1)$ are sufficiently well localized states in position, far from each other in order to prevent any overlap. Thus, we can consider them as distinguishable (in a macroscopic sense), and we approximate $\langle\alpha|\beta\rangle \simeq 0$. The second system (S2) will serve as a point-like probe of the gravitational field generated by S1, it has mass m_2 and state $\phi(r_2)$. The state $\phi(r_2)$ is initially assumed to be localized in position and centered along the y direction [cf. Fig. 8]. The question we address is: which is the gravitational field, generated by the quantum superposition of S1, that S2 experiences? We probe the following two different scenarios.

Quantum Gravity Scenario. Although we do not have a quantum theory of gravity so far, one can safely claim that, regardless of how it is realized, it would manifest in S1 generating a superposition of gravitational fields. As discussed in the introduction, the assessment of this property precedes the quest to ascertain the existence of the graviton and the characterization of its properties, at least as far as the static, low-energy, non-relativistic regime we are considering is concerned. Linearity is the very characteristic trait of quantum theory, and one expects it to be preserved by any quantum theory of gravity.

The reaction of S2 is then to go in a superposition of being attracted towards the region where $|\alpha\rangle$ sits and where $|\beta\rangle$ does. The final two-body state will have the following entangled form

$$\Psi_{\text{QG}}^{\text{final}}(r_1, r_2) = \frac{\alpha(\mathbf{r}_1)\phi_\alpha(\mathbf{r}_2) + \beta(\mathbf{r}_1)\phi_\beta(\mathbf{r}_2)}{\sqrt{2}}, \tag{9}$$

where $\phi_\alpha(\mathbf{r}_2)$ ($\phi_\beta(\mathbf{r}_2)$) represents the state of S2 attracted towards the region where $|\alpha\rangle$ ($|\beta\rangle$) rests. The motion in each branch of the superposition is produced by the potential

$$\hat{V}_\gamma(\hat{r}_2) = -Gm_2 \int \mathrm{d}r_1 \frac{\rho_\gamma(\mathbf{r}_1)}{|r_1 - \hat{r}_2|}, \qquad (\gamma = \alpha, \beta). \tag{10}$$

where $\rho_\gamma(\mathbf{r}_1)$ is the mass density of S1, centred in $\langle\hat{r}_1\rangle_\gamma = \langle\gamma|\hat{r}_1|\gamma\rangle$. We assume that S1 does not move appreciably during the time of the experiment (also quantum fluctuations can be neglected); clearly, such a situation can be assumed only as long as the S1 superposition lives. We further assume that its mass density is essentially spheric, so that the gravitational interaction can be approximated by

$$\hat{V}_\gamma(\hat{r}_2) \approx -\frac{Gm_1m_2}{|\langle\hat{\mathbf{r}}_1\rangle_\gamma - \hat{\mathbf{r}}_2|}, \quad (\gamma = \alpha, \beta). \tag{11}$$

Semiclassical Gravity Scenario. The second scenario sees gravity as fundamentally classical. In this case, it is not clear which characteristics one should expect from the gravitational field generated by a superposition. However, in analogy with classical mechanics, one can assume that is the mass density $\rho(\mathbf{r}_1) = (\rho_\alpha(\mathbf{r}_1) + \rho_\beta(\mathbf{r}_1))/2$ of the system in superposition that produces the gravitational field. This is also what is predicted by the Schrödinger-Newton equation (see Sect. 3). The final two-body state will be of the form

$$\Psi_{\text{CG}}^{\text{final}}(r_1, r_2) = \frac{\alpha(\mathbf{r}_1) + \beta(\mathbf{r}_1)}{\sqrt{2}}\phi(r_2), \tag{12}$$

where the difference with Eq. (9) is clear. The gravitational potential becomes

$$\hat{V}_{\text{cl}}(\hat{r}_2) \approx \tfrac{1}{2} \sum_{\gamma=\alpha,\beta} \hat{V}_\gamma(\hat{r}_2), \tag{13}$$

where $\hat{V}_\gamma(\hat{r}_2)$ can be eventually approximated as in Eq. (11).

Experimental progress with levitated mechanical systems makes is possible to reach a parameter regime to experimentally resolve the difference between the quantum and semiclassical scenarios as shown in our paper [98]. Other interferometric [100, 101] and non-interferometric [102] tests of the nature of gravity have been proposed. They are based on the detection of entanglement between two probes, respectively coupled to two different massive systems, which interact through gravity (NV center spins for [100] and cavity fields for [102]). Clearly, to have such entanglement, each of the three couples of interconnected systems (probe 1, system 1, system 2 and probe 2) there considered needs to be entangled on their own. Moreover, the entanglement between the two massive systems is inevitably small due to its gravitational nature. Conversely, our proposal benefits from having only a single massive system involved in the interconnection, which reduces correlation losses. In addition, we provide a second method for discerning the nature of gravity: the individuation of a second peak in the DNS. The latter does not rely on delicate measurements of quantum correlations but can be assessed through standard optomechanical detection schemes.

3.4 Concluding Remarks on Testing the Interplay of Quantum Mechanics and Gravity in the Low Energy Regime

While matterwave interferometer experiments have been performed in the low mass regime, see Fig. 4, the higher mass range, all the way up to milligram masses is unexplored by any experiment and especially not by any quantum experiment. Optomechanical devices and especially levitated particles are able to bridge this enormous mass gap; being in a quantum mechanical state and very massive at the same time. Levitated mechanical systems hold promise to test new physics in that new mass range. A variety of theoretical proposals and ideas for the interplay between quantum mechanics and gravity will become testable in this very mass range. The study of gravitational decoherence, the Schrödinger-Newton equation and the gravity of a quantum state provide concrete routes for experimental exploration.

4 Simulation of the Stern Gerlach Experiment Using Wigner Functions

The Stern-Gerlach (SG) experiment [103] is a seminal example of a quantum experiment involving coupling between internal and external degrees of freedom. In this experiment, an electron or nuclear spin interacts with a spatially inhomogeneous magnetic field through the magnetic Zeeman interaction. The outcome of the Stern-Gerlach experiment is, of course, "well-known": an incident molecular beam of particles with spin-1/2 is separated by the inhomogeneous magnetic field into two beams, each corresponding to particles with well-defined spin angular momenta along the field direction. But how does this separation happen *in detail*, on the level of the spatial quantum state?

In a recent article [104] we used an *extended Wigner function* (EWF) which includes the presence of internal degrees of freedom in the propagating particle, and the coupling of those internal degrees of freedom to inhomogeneous external fields (Fig. 9).

The Wigner function $W(x, p)$ is a joint quasi-probability density function defined over the combined domains of the spatial coordinate(s) x and its associated momentum (momenta) p. It is defined as a Weyl integral transform of the density operator $\hat{\rho} = |\psi\rangle\langle\psi|$, of the following form:

$$W(x, p) = \frac{1}{h} \int e^{-\frac{ips}{\hbar}} \langle x + \tfrac{s}{2}|\hat{\rho}|x - \tfrac{s}{2}\rangle \, ds. \tag{14}$$

Consider a particle with a finite number of internal quantum states. In the discussion below, we refer to these internal states as "spin states", although the same formalism applies to non-spin degrees of freedom, such as quantized rotational and

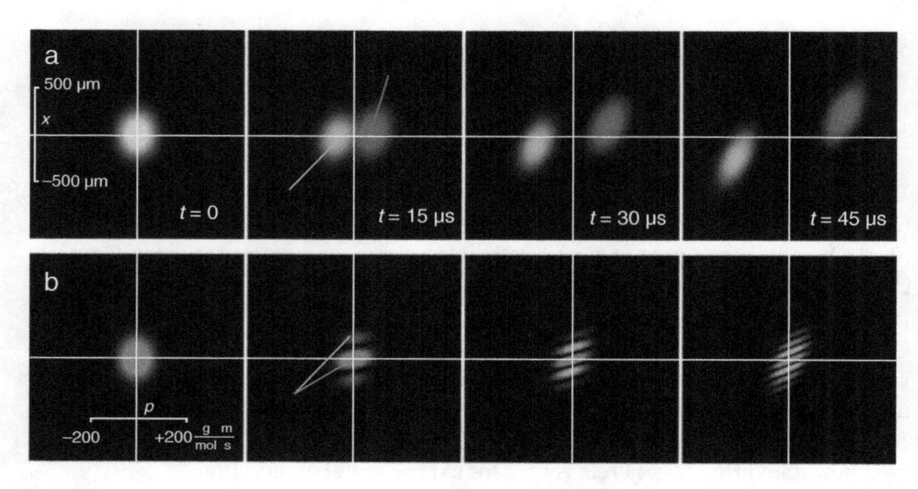

Fig. 9 **a** Evolution of $W_{\alpha\alpha}$ and $W_{\beta\beta}$ under the influence of a magnetic field gradient in a Stern-Gerlach experiment on *Ag* atoms in a field gradient of 10 G μm^{-1}, moving at a velocity of 550 m/s (rms velocity at an oven temperature of 1300 K). **b** Evolution of the real part of the off-diagonal element $W_{\alpha\beta}$, assuming a coherent state initially polarised along the x-axis. The strength of the magnetic field gradient has been reduced by a factor of 5×10^4 compared to A in order to make the spatial modulation visible. The shearing of the fine structure of the Wigner function represents decoherence. Figure and capture taken from Ref. [104]

vibrational states. We extend the Wigner function by combining it with the density operator formalism commonly used in the quantum description of magnetic resonance. The definition of the Wigner function is extended by projecting the density operator onto the spin-state specific position state $|x, \eta\rangle$, where $\eta = \alpha, \beta, \ldots$ denotes the spin state. This results in a Wigner probability density matrix $W_{\eta\xi}(x, p)$, whose elements depend parametrically on the positional variables and their associated momenta:

$$W_{\eta\xi}(x, p) = \frac{1}{h} \int e^{-\frac{ips}{\hbar}} \langle x + \tfrac{s}{2}, \eta|\hat{\rho}|x - \tfrac{s}{2}, \xi\rangle \, ds. \tag{15}$$

This means the extended Wigner function can be used to directly simulate the SG experiment. In the Stern-Gerlach experiment, a beam of spin-1/2 particles is exposed to a lateral magnetic field gradient. We define the axis of the molecular beam apparatus as z, and assume that the magnetic field varies in the transverse x-direction. The potential energy part of the Hamiltonian in the presence of an external magnetic field $\mathbf{B}$ is then given by

$$U(\hat{\mathbf{S}}, x) = -\hbar\gamma \mathbf{B}(\mathbf{x}) \cdot \hat{\mathbf{S}}. \tag{16}$$

The original magnet design used by Stern and Gerlach [103] produces divergent magnetic field lines at the location of the beam. This corresponds to a biaxial magnetic field gradient tensor, requiring two spatial dimensions to be included in the Wigner function. To avoid this complication, we use a different arrangement, in which the magnetic field gradient is uniaxial. In this case, the magnetic field lines are all parallel, but vary in density in the direction perpendicular to the magnetic field itself. Magnetic fields of this type occur in quadrupole polarisers.

We assume the magnetic field points along the y-axis, and varies linearly in magnitude along the x-axis, $\mathbf{B}(\mathbf{x},\mathbf{y},\mathbf{z}) = \left(\mathbf{B_{y0}} + \mathbf{x}\,\mathbf{G_{xy}}\right)\,\mathbf{e_y}$, where B_{y0} is the magnetic field at $x = 0$, and $G_{xy} = \partial B_y/\partial x$. This field is fully consistent with Maxwell's equations, since it satisfies $\nabla \cdot \mathbf{B} = \mathbf{0}$. The field gradient has only a single non-zero cartesian component $\nabla\mathbf{B} = \mathbf{G_{xy}}\ \mathbf{e_x}\mathbf{e_y}$. We choose the spin states $|\alpha\rangle$ and $|\beta\rangle$ as the eigenstates of $\hat{S}_y$, such that the matrix elements of the potential part of the Hamiltonian are

$$\begin{array}{ll} U_{\alpha\alpha}(x) = -\frac{\gamma\hbar}{2}B_y(x) & U_{\alpha\beta}(x) = 0 \\ U_{\beta\alpha}(x) = 0 & U_{\beta\beta}(x) = +\frac{\gamma\hbar}{2}B_y(x). \end{array} \tag{17}$$

The resulting equations of motion for the EWF matrix elements are given in the SI.

In its original form, the Stern-Gerlach experiment was conducted on a beam of Ag atoms emanating from an oven at a temperature of about 1300 K. The magnetic field gradient was of the order of 10 G/cm over a length of 3.5 cm [105]. For simplicity, we ignore the nuclear spin of Ag, and treat the atoms as (electron) spin 1/2 particles. In the case of magnetic fields larger than the hyperfine splitting (about 610 G in the case of Ag), this is a good approximation, since the nuclear and the electron spin states are essentially decoupled. The root mean square velocity of Ag atoms 1300 K is approximately 550 m/s. After leaving the oven, the Ag atoms are collimated by a pair of collimation slits 30 μm wide and separated by 3 cm. The longitudinal momentum of the silver atoms is approximately 6×10^4 g mol^{-1} ms^{-1}. The collimation aspect ratio of 1:1000 therefore results in a transverse momentum uncertainty of $\Delta p = 60$ g mol^{-1} ms^{-1}, which corresponds to a 30 μm wide beam with a transverse coherence length of about $l_c = h/\Delta p \approx 7$ nm.

An unpolarised beam entering the magnetic field gradient is represented by a unity spin density matrix, such that $W_{\alpha\alpha}(t = 0) = W_{\beta\beta}(t = 0) = W_0(x, p)$, where the initial state $W_0(x, p)$ is a two-dimensional normalised Gaussian function centred at $(x, p) = (0, 0)$, with widths given by coherence length l_c and the beam width Δx (cf. SI). The off-diagonal Wigner functions vanish: $W_{\alpha\beta} = W_{\beta\alpha} \equiv 0$, and the diagonal ones can be obtained in closed form by integrating the equations of motion (cf. SI).

In conclusion, the SG magnet can be used as coherent beam splitter, but the original experiment did not do the recombination or any other protocol to demonstrate the quantum correlation. When a coherent superposition spin state is provided at the SG input then a coherent spatial superposition of the centre of mass motion of the particle can be achieved. This has been finally demonstrated by the group of Ron Folman [106] for the case of atom interferometry and is used as central ingredient for a recent proposal of the generation of macroscopic quantum superposition [100, 101].

Acknowledgements I would like to thank my research team members at Southampton and external collaborators for joint work on testing fundamental physics with mechanical systems in table-top experiments over the past ten years. We are getting there.

References

1. J.P. Toennies, H. Schmidt-Böcking, B. Friedrich, J.C. Lower, Otto Stern (1888–1969): The founding father of experimental atomic physics. Ann. Phys. **523**(12), 1045–1070 (2011)
2. S. Deachapunya, P.J. Fagan, A.G. Major, E. Reiger, H. Ritsch, A. Stefanov, H. Ulbricht, M. Arndt, Slow beams of massive molecules. Euro. Phys. J. D **46**(2), 307–313 (2008)
3. P. Asenbaum, S. Kuhn, S. Nimmrichter, U. Sezer, M. Arndt, Cavity cooling of free silicon nanoparticles in high vacuum. Nat. Commun. **4**(1), 1–7 (2013)
4. J. Millen, T.S. Monteiro, R. Pettit, A.N. Vamivakas, Optomechanics with levitated particles. Rep. Prog. Phys. **83**(2), 026401 (2020)
5. A. Ashkin, Optical trapping and manipulation of neutral particles using lasers: A reprint volume with commentaries (2006)
6. J. Millen, P.Z.G. Fonseca, T. Mavrogordatos, T.S. Monteiro, P.F. Barker, Cavity cooling a single charged levitated nanosphere. Phys. Rev. Lett. **114**(12), 123602 (2015)
7. D.C. Moore, A.D. Rider, G. Gratta, Search for millicharged particles using optically levitated microspheres. Phys. Rev. Lett. **113**(25), 251801 (2014)
8. C. Timberlake, G. Gasbarri, A. Vinante, A. Setter, H. Ulbricht, Acceleration sensing with magnetically levitated oscillators above a superconductor. Appl. Phys. Lett. **115**(22), 224101 (2019)
9. Y.Y. Fein, P. Geyer, P. Zwick, F. Kialka, S. Pedalino, M. Mayor, S. Gerlich, M. Arndt, Quantum superposition of molecules beyond 25 kDa. Nat. Phys. **15**(12), 1242–1245 (2019)
10. S. Nimmrichter, K. Hornberger, Macroscopicity of mechanical quantum superposition states. Phys. Rev. Lett. **110**(16), 160403 (2013)
11. A. Bassi, G. Ghirardi, Dynamical reduction models. Phys. Rep. **379**(5–6), 257–426 (2003)
12. G.C. Ghirardi, A. Rimini, T. Weber, Unified dynamics for microscopic and macroscopic systems. Phys. Rev. D **34**(2), 470 (1986)
13. G.C. Ghirardi, P. Pearle, A. Rimini, Markov processes in Hilbert space and continuous spontaneous localization of systems of identical particles. Phys. Rev. A **42**(1), 78 (1990)
14. A. Bassi, K. Lochan, S. Satin, T.P. Singh, H. Ulbricht, Models of wave-function collapse, underlying theories, and experimental tests. Rev. Mod. Phys. **85**(2), 471 (2013)
15. S.L. Adler, Lower and upper bounds on CSL parameters from latent image formation and IGM heating. J. Phys. A: Math. Theor. **40**(12), 2935 (2007)
16. S. Eibenberger, S. Gerlich, M. Arndt, M. Mayor, J. Tüxen, Matter–wave interference of particles selected from a molecular library with masses exceeding 10000 amu. Phys. Chem. Chem. Phys. **15**(35), 14696–14700 (2013)
17. T. Kovachy, P. Asenbaum, C. Overstreet, C.A. Donnelly, S.M. Dickerson, A. Sugarbaker, J.M. Hogan, M.A. Kasevich, Quantum superposition at the half-metre scale. Nature **528**(7583), 530–533 (2015)
18. K.C. Lee, M.R. Sprague, B.J. Sussman, J. Nunn, N.K. Langford, X.M. Jin, T. Champion, P. Michelberger, K.F. Reim, D. England, D. Jaksch, Entangling macroscopic diamonds at room temperature. Science **334**(6060), 1253–1256 (2011)
19. M. Armano, H. Audley, J. Baird, P. Binetruy, M. Born, D. Bortoluzzi, E. Castelli, A. Cavalleri, A. Cesarini, A.M. Cruise, K. Danzmann, Beyond the required LISA free-fall performance: new LISA Pathfinder results down to 20 Hz. Phys. Rev. Lett. **120**(6), 061101 (2018)
20. M. Armano, H. Audley, G. Auger, J.T. Baird, M. Bassan, P. Binetruy, M. Born, D. Bortoluzzi, N. Brandt, M. Caleno, L. Carbone, Sub-femto-g free fall for space-based gravitational wave observatories: LISA pathfinder results. Phys. Rev. Lett. **116**(23), 231101 (2016)
21. T. Kovachy, J.M. Hogan, A. Sugarbaker, S.M. Dickerson, C.A. Donnelly, C. Overstreet, M.A. Kasevich, Matter wave lensing to picokelvin temperatures. Phys. Rev. Lett. **114**(14), 143004 (2015)

22. S.L. Adler, A. Vinante, Bulk heating effects as tests for collapse models. Phys. Rev. A **97**(5), 052119 (2018)
23. K. Piscicchia, A. Bassi, C. Curceanu, R.D. Grande, S. Donadi, B.C. Hiesmayr, A. Pichler, CSL collapse model mapped with the spontaneous radiation. Entropy **19**(7), 319 (2017)
24. A. Vinante, M. Bahrami, A. Bassi, O. Usenko, G.H.C.J. Wijts, T.H. Oosterkamp, Upper bounds on spontaneous wave-function collapse models using millikelvin-cooled nanocantilevers. Phys. Rev. Lett. **116**(9), 090402 (2016)
25. M. Torol, G. Gasbarri, A. Bassi, Colored and dissipative continuous spontaneous localization model and bounds from matter-wave interferometry. Phys. Lett. A **381**(47), 3921–3927 (2017)
26. M. Carlesso, A. Bassi, Current tests of collapse models: How far can we push the limits of quantum mechanics? in *Quantum Information and Measurement* (Optical Society of America, 2019, April), pp. S1C-3
27. M. Toroš, A. Bassi, Bounds on quantum collapse models from matter-wave interferometry: Calculational details. J. Phys. A: Math. Theoretical **51**(11), 115302 (2018)
28. C. Gardiner, P. Zoller, *Quantum Noise: A Handbook of Markovian and Non-Markovian Quantum Stochastic Methods with Applications to Quantum Optics* Vol. 56 (Springer Science and Business Media, 2004)
29. H.P. Breuer, F. Petruccione, *The Theory of Open Quantum Systems* (Oxford University Press, Oxford, 2002)
30. J. Bateman, S. Nimmrichter, K. Hornberger, H. Ulbricht, Near-field interferometry of a free-falling nanoparticle from a point-like source. Nat. Commun. **5**(1), 1–5 (2014)
31. H. Pino, J. Prat-Camps, K. Sinha, B.P. Venkatesh, O. Romero-Isart, On-chip quantum interference of a superconducting microsphere. Quantum Sci. Technol. **3**(2), 025001 (2018)
32. C. Wan, M. Scala, G.W. Morley, A.A. Rahman, H. Ulbricht, J. Bateman, P.F. Barker, S. Bose, M.S. Kim, Free nano-object Ramsey interferometry for large quantum superpositions. Phys. Rev. Lett. **117**(14), 143003 (2016)
33. R. Kaltenbaek, M. Aspelmeyer, P.F. Barker, A. Bassi, J. Bateman, K. Bongs, S. Bose, C. Braxmaier, Brukner, Christophe, B., Chwalla, M., Macroscopic quantum resonators (MAQRO): 2015 update. EPJ Quantum Technol. **3**(1), 5 (2016)
34. M. Bahrami, A. Smirne, A. Bassi, Role of gravity in the collapse of a wave function: A probe into the diósi-penrose model. Phys. Rev. A **90**(6), 062105 (2014)
35. B. Collett, P. Pearle, Wavefunction collapse and random walk. Found. Phys. **33**(10), 1495–1541 (2003)
36. S.L. Adler, Stochastic collapse and decoherence of a non-dissipative forced harmonic oscillator. J. Phys. A: Math. Gen. **38**(12), 2729 (2005)
37. K. Hornberger, S. Gerlich, P. Haslinger, S. Nimmrichter, M. Arndt, Colloquium: Quantum interference of clusters and molecules. Rev. Mod. Phys. **84**(1), 157 (2012)
38. S. Nimmrichter, K. Hornberger, K. Hammerer, Optomechanical sensing of spontaneous wave-function collapse. Phys. Rev. Lett. **113**(2), 020405 (2014)
39. S. Bera, B. Motwani, T.P. Singh, H. Ulbricht, A proposal for the experimental detection of CSL induced random walk. Sci. Rep. **5**(1), 1–10 (2015)
40. L. Diósi, Testing spontaneous wave-function collapse models on classical mechanical oscillators. Phys. Rev. Lett. **114**(5), 050403 (2015)
41. D. Goldwater, M. Paternostro, P.F. Barker, Testing wave-function-collapse models using parametric heating of a trapped nanosphere. Phys. Rev. A **94**(1), 010104 (2016)
42. M. Carlesso, M. Paternostro, H. Ulbricht, A. Vinante, A. Bassi, Non-interferometric test of the continuous spontaneous localization model based on rotational optomechanics. New J. Phys. **20**(8), 083022 (2018)
43. M. Carlesso, A. Bassi, P. Falferi, A. Vinante, Experimental bounds on collapse models from gravitational wave detectors. Phys. Rev. D **94**(12), 124036 (2016)
44. W. Paul, Electromagnetic traps for charged and neutral particles. Rev. Mod. Phys. **62**(3), 531 (1990)
45. A. Vinante, A. Pontin, M. Rashid, M. Torol, P.F. Barker, H. Ulbricht, Testing collapse models with levitated nanoparticles: Detection challenge. Phys. Rev. A **100**(1), 012119 (2019)

46. R. Mishra, A. Vinante, T.P. Singh, Testing spontaneous collapse through bulk heating experiments: An estimate of the background noise. Phys. Rev. A **98**(5), 052121 (2018)
47. M. Bilardello, S. Donadi, A. Vinante, A. Bassi, Bounds on collapse models from cold-atom experiments. Phys. A **462**, 764–782 (2016)
48. A. Pontin, N.P. Bullier, M. Torol, P.F. Barker, Ultranarrow-linewidth levitated nano-oscillator for testing dissipative wave-function collapse. Phys. Rev. Res. **2**(2), 023349 (2020)
49. B.R. Slezak, C.W. Lewandowski, J.F. Hsu, D'Urso, B., Cooling the motion of a silica microsphere in a magneto-gravitational trap in ultra-high vacuum. New J. Phys. **20**(6), 063028 (2018)
50. D. Zheng, Y. Leng, X. Kong, R. Li, Z. Wang, X. Luo, J. Zhao, C.K. Duan, P. Huang , J. Du, Room temperature test of wave-function collapse using a levitated micro-oscillator (2019). arXiv preprint arXiv:1907.06896
51. O. Romero-Isart, L. Clemente, C. Navau, A. Sanchez, J.I. Cirac, Quantum magnetomechanics with levitating superconducting microspheres. Phys. Rev. Lett. **109**(14), 147205 (2012)
52. B. van Waarde, *The lead zeppelin: a force sensor without a handle*, Ph.D. Thesis, Leiden University (2016)
53. A. Vinante, P. Falferi, G. Gasbarri, A. Setter, C. Timberlake, H. Ulbricht, Ultralow mechanical damping with Meissner-levitated ferromagnetic microparticles. Phys. Rev. Appl. **13**(6), 064027 (2020)
54. J. Prat-Camps, C. Teo, C.C. Rusconi, W. Wieczorek, O. Romero-Isart, Ultrasensitive inertial and force sensors with diamagnetically levitated magnets. Phys. Rev. Appl. **8**(3), 034002 (2017)
55. M. Bahrami, A. Bassi, H. Ulbricht, Testing the quantum superposition principle in the frequency domain. Phys. Rev. A **89**(3), 032127 (2014)
56. S. Sturm, F. Köhler, J. Zatorski, A. Wagner, Z. Harman, G. Werth, W. Quint, C.H. Keitel, K. Blaum, High-precision measurement of the atomic mass of the electron. Nature **506**(7489), 467–470 (2014)
57. M. Weitz, A. Huber, F. Schmidt-Kaler, D. Leibfried, T.W. Hänsch, Precision measurement of the hydrogen and deuterium 1 S ground state Lamb shift. Phys. Rev. Lett. **72**(3), 328 (1994)
58. E. Oelker, R.B. Hutson, C.J. Kennedy, L. Sonderhouse, T. Bothwell, A. Goban, D. Kedar, C. Sanner, J.M. Robinson, G.E. Marti , D.G. Matei, Demonstration of 4.8×10^{-17} stability at 1 s for two independent optical clocks. Nat. Photon. 13(10), pp.714-719 (2019)
59. R. Penrose, On the gravitization of quantum mechanics 1: Quantum state reduction. Found. Phys. **44**(5), 557–575 (2014)
60. L. Rosenfeld, On quantization of fields. Nuclear Phys. **40**, 353–356 (1963)
61. S. Carlip, Is quantum gravity necessary? Class. Quantum Gravity **25**(15), 154010 (2008)
62. C. Møller, Les théories relativistes de la gravitation. Colloques Internationaux CNRS **91**(1) (1962)
63. D.N. Page, C.D. Geilker, Indirect evidence for quantum gravity. Phys. Rev. Lett. **47**(14), 979 (1981)
64. K. Eppley, E. Hannah, The necessity of quantizing the gravitational field. Found. Phys. **7**(1–2), 51–68 (1977)
65. N. Gisin, Stochastic quantum dynamics and relativity. Helv. Phys. Acta **62**(4), 363–371 (1989)
66. J. Mattingly, Why Eppley and Hannah's thought experiment fails. Phys. Rev. D **73**(6), 064025 (2006)
67. C. Kiefer, Why quantum gravity? in *Approaches to Fundamental Physics* (Springer, Berlin, Heidelberg, 2007), pp. 123–130
68. M. Albers, C. Kiefer, M. Reginatto, Measurement analysis and quantum gravity. Phys. Rev. D **78**(6), 064051 (2008)
69. D. Giulini, A. Großardt, Gravitationally induced inhibitions of dispersion according to the Schrödinger–Newton equation. Class. Quantum Gravity **28**(19), 195026 (2011)
70. D. Giulini, A. Großardt, Gravitationally induced inhibitions of dispersion according to a modified Schrödinger–Newton equation for a homogeneous-sphere potential. Class. Quantum Gravity **30**(15), 155018 (2013)

71. L. Diósi, Gravitation and quantum-mechanical localization of macro-objects. Phys. Lett. A **105**(4–5), 199–202 (1984)
72. R. Penrose, Quantum computation, entanglement and state reduction. Philos. Trans. R. Soc. London. Ser. A: Math. Phys. Eng. Sci. **356**(1743), 1927–1939
73. H. Yang, H. Miao, D.S. Lee, B. Helou, Y. Chen, Macroscopic quantum mechanics in a classical spacetime. Phys. Rev. Lett. **110**(17), 170401 (2013)
74. A. Großardt, J. Bateman, H. Ulbricht, A. Bassi, Optomechanical test of the Schrödinger-Newton equation. Phys. Rev. D **93**(9), 096003 (2016)
75. R. Bekenstein, R. Schley, M. Mutzafi, C. Rotschild, M. Segev, Optical simulations of gravitational effects in the Newton–Schrödinger system. Nat. Phys. **11**(10), 872–878 (2015)
76. T. Roger, C. Maitland, K. Wilson, N. Westerberg, D. Vocke, E.M. Wright, D. Faccio, Optical analogues of the Newton–Schrödinger equation and boson star evolution. Nat. Commun. **7**(1), 1–8 (2016)
77. M. Zych, F. Costa, I. Pikovski, T.C. Ralph, Č. Brukner, General relativistic effects in quantum interference of photons. Class. Quantum Gravity **29**(22), 224010 (2012)
78. M. Zych, F. Costa, I. Pikovski, Č. Brukner, Quantum interferometric visibility as a witness of general relativistic proper time. Nat. Commun. **2**, 505 (2011)
79. I. Pikovski, M. Zych, F. Costa, Č. Brukner, Universal decoherence due to gravitational time dilation. Nat. Phys. **11**(8), 668–672 (2015)
80. M. Torol, A. Bassi, Bounds on quantum collapse models from matter-wave interferometry: Calculational details. J. Phys. A: Math. Theor. **51**(11), 115302 (2018)
81. Y. Margalit, Z. Zhou, S. Machluf, D. Rohrlich, Y. Japha, R. Folman, A self-interfering clock as a "which path" witness. Science **349**(6253), 1205–1208 (2015)
82. S. Nimmrichter, K. Hornberger, Stochastic extensions of the regularized Schrödinger-Newton equation. Phys. Rev. D **91**(2), 024016 (2015)
83. S. Bera, S. Donadi, K. Lochan, T.P. Singh, A comparison between models of gravity induced decoherence. Found. Phys. **45**(12), 1537–1560 (2015)
84. G. Gasbarri, M. Torol, S. Donadi, A. Bassi, Gravity induced wave function collapse. Phys. Rev. D **96**(10), 104013 (2017)
85. O. Romero-Isart, A.C. Pflanzer, F. Blaser, R. Kaltenbaek, N. Kiesel, M. Aspelmeyer, J.I. Cirac, Large quantum superpositions and interference of massive nanometer-sized objects. Phys. Rev. Lett. **107**(2), 020405 (2011)
86. L. Diosi, A universal master equation for the gravitational violation of quantum mechanics. Phys. Lett. A **120**(8), 377–381 (1987)
87. R. Penrose, On gravity's role in quantum state reduction. Gen. Relativ. Gravit. **28**(5), 581–600 (1996)
88. S. Bose, K. Jacobs, P.L. Knight, Preparation of nonclassical states in cavities with a moving mirror. Phys. Rev. A **56**(5), 4175 (1997)
89. S. Bose, K. Jacobs, P.L. Knight, Scheme to probe the decoherence of a macroscopic object. Phys. Rev. A **59**(5), 3204 (1999)
90. M.R. Vanner, I. Pikovski, G.D. Cole, M.S. Kim, Č. Brukner, K. Hammerer, G.J. Milburn, M. Aspelmeyer, Pulsed quantum optomechanics. Proc. National Acad. Sci. **108**(39), 16182–16187 (2011)
91. W. Marshall, C. Simon, R. Penrose, D. Bouwmeester, Towards quantum superpositions of a mirror. Phys. Rev. Lett. **91**(13), 130401 (2003)
92. K. Hornberger, S. Uttenthaler, B. Brezger, L. Hackermüller, M. Arndt, A. Zeilinger, Collisional decoherence observed in matter wave interferometry. Phys. Rev. Lett. **90**(16), 160401 (2003)
93. S.L. Adler, Gravitation and the noise needed in objective reduction models Quantum Nonlocality and Reality: 50 Years of Bell's Theorem ed M Bell and S Gao (2016)
94. M. Bronstein, Republication of: Quantum theory of weak gravitational fields. Gen. Relativ. Gravit. **44**(1), 267–283 (2012)
95. B.S. DeWitt, D. Bryce Seligman, *The Global Approach to Quantum Field Theory* (Vol. 114) (Oxford University Press, USA, 2003)
96. A. Peres, D.R. Terno, Hybrid classical-quantum dynamics. Phys. Rev. A **63**(2), 022101 (2001)

97. C. Marletto, V. Vedral, Why we need to quantise everything, including gravity. npj Quantum Inf. 3(1), 1–5 (2017)
98. M. Carlesso, A. Bassi, M. Paternostro, H. Ulbricht, Testing the gravitational field generated by a quantum superposition. New J. Phys. **21**(9), 093052 (2019)
99. M. Carlesso, M. Paternostro, H. Ulbricht, A. Bassi, When Cavendish meets Feynman: A quantum torsion balance for testing the quantumness of gravity (2017). arXiv preprint arXiv:1710.08695
100. S. Bose, A. Mazumdar, G.W. Morley, H. Ulbricht, M. Torol, M. Paternostro, A.A. Geraci, P.F. Barker, M.S. Kim, G. Milburn, Spin entanglement witness for quantum gravity. Phys. Rev. Lett. **119**(24), 240401 (2017)
101. C. Marletto, V. Vedral, Gravitationally induced entanglement between two massive particles is sufficient evidence of quantum effects in gravity. Phys. Rev. Lett. **119**(24), 240402 (2017)
102. H. Miao, D. Martynov, H. Yang, A. Datta, Quantum correlations of light mediated by gravity. Phys. Rev. A **101**(6), 063804 (2020)
103. W. Gerlach, O. Stern, Der experimentelle Nachweis des magnetischen Moments des Silberatoms. ZPhy **8**(1), 110–111 (1922)
104. M. Utz, M.H. Levitt, N. Cooper, H. Ulbricht, Visualisation of quantum evolution in the Stern–Gerlach and Rabi experiments. Phys. Chem. Chem. Phys. **17**(5), 3867–3872 (2015)
105. B. Friedrich, D. Herschbach, Stern and Gerlach: How a bad cigar helped reorient atomic physics. Phys. Today **56**(12), 53–59 (2003)
106. S. Machluf, Y. Japha, R. Folman, Coherent Stern–Gerlach momentum splitting on an atom chip. Nat. Commun. **4**(1), 1–9 (2013)

8

Quantum Effects in Cold and Controlled Molecular Dynamics

Christiane P. Koch

Abstract This chapter discusses three examples of quantum effects that can be observed in state-of-the-art experiments with molecular beams—scattering resonances as a probe of interparticle interactions in cold collisions, the protection of Fano-Feshbach resonances against decay despite resonant coupling to a scattering continuum, and a circular dichroism in photoelectron angular distributions arising in the photoionization of randomly oriented chiral molecules. The molecular beam setup provides molecules in well-defined quantum states. This, together with a theoretical description based on first principles, allows for excellent agreement between theoretical prediction and experimental observation and thus a rigorous understanding of the observed quantum effects.

1 Introduction

When you ask young students entering a university physics course today for the term they associate most with quantum mechanics, many of them will respond with "entanglement". This reflects the rise of quantum information science out of an often ridiculed ivory tower to the decision-making levels of the big tech companies and to the headlines of well-respected media outlets. In the waves of excitement created by the "second quantum revolution" [1, 2], it may be overlooked that features more traditionally associated with quantum mechanics continue to fascinate and challenge our classically trained intuition.

This chapter reviews three examples of such quantum effects beyond entanglement from recent work of my group—tunneling resonances that emerge in cold collisions and that can be used to probe interparticle interactions [3], Fano-Feshbach resonances that can be protected against decay by a suitable phase condition [4], and quantum pathway interference in the circular dichroism of photoelectrons that is observed after the photoionization of chiral molecules [5, 6]. All three examples share a rigorous theoretical description based on first principles. More importantly

C. P. Koch (✉)
Arnimallee 14, 14195 Berlin, Germany
e-mail: christiane.koch@fu-berlin.de

for the present contribution, the quantum effects discussed here have been observed in experiments with molecular beams [3, 4, 7]. They thus testify to the topicality and continuing significance of the molecular beam technique developed by Otto Stern and colleagues.

2 Quantum Scattering Resonances in Cold Collisions

The wave nature of colliding particles emerges most prominently at low scattering energies [8, 9], where quantum resonances dominate the scattering cross section before threshold laws take over [10]. While quantum resonances are also present at higher scattering energies, this presence is hidden in the ensemble average over all quantum states that are populated at a given energy. In other words, collisions are "cold" when only a few partial waves contribute to the scattering cross section [8]. At the corresponding collision energies, the dynamics are often dominated by the long-range behavior of the interparticle interactions [10], easing the interpretation of the collision studies.

In order to observe quantum scattering resonances experimentally, one needs to ensure a sufficiently narrow velocity distribution—narrower than the resonance width—in addition to the capability to finely tune the relative kinetic energy of the colliding particles to very low values. Both requirements can be met with merged neutral beams [11, 12], the use of which has allowed for the observation of tunneling resonances in Penning ionization reactions [13]. Since then, scattering resonances have also been observed in inelastic low-energy collisions, see Ref. [14] and references therein.

Penning ionization reactions occur when the excitation energy of a particle prepared in a metastable quantum state is sufficient to ionize its collision partner [15]. Metastable nobel gas atoms, for example, feature excitation energies of more than 10 eV which is above the ionization potential of most molecules. The scattering resonances in this case are tunneling, or shape, resonances that occcur when the kinetic energy of the colliding particles matches the energy of a quasi-bound state that is trapped behind the rotational barrier of an otherwise barrier-less potential, cf. Fig. 1. Peaks in the ionization cross section indicate not only the presence of a tunneling resonance but also highlight the corresponding quantization of the intermolecular motion. In a more pictorial description, when the colliding particles hit a resonance upon tunneling through the rotational barrier, their amplitude gets trapped at short interparticle distances. This results in peaks in the cross section since the ionization probability grows inverse-exponentially with interparticle distance [15].

The experiments leading to the first observation of tunneling resonances in Penning ionization reactions were carried out with metastable helium colliding with argon atoms and dihydrogen molecules [13]. At the time, no qualitative difference was observed for the Penning ionization of an atom compared to that of a molecule. In general, however, one would expect the molecular degrees of freedom—rotations or vibrations—as well as the anisotropy of the interparticle interaction potential (the

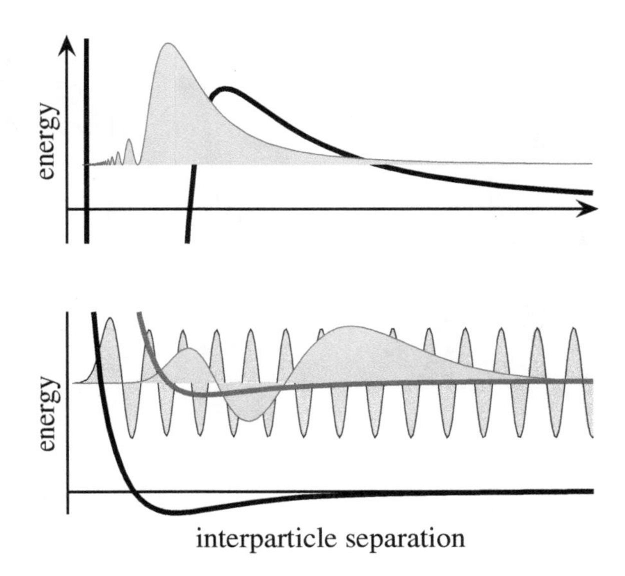

Fig. 1 Tunneling, or shape, resonances (top) versus Fano-Feshbach resonances (bottom): Tunneling resonances form on a single potential curve displaying a barrier. In contrast, Fano-Feshbach resonances arise from the coupling of scattering states to a bound state belonging to another scattering channel which is energetically closed and asymptotically characterized by a different set of quantum numbers. In both cases, scattering amplitude gets trapped at short interparticle distance

difference between the molecule hitting the metastable atom head-on or in a T-shaped geometry) to come into play and modify the cross section, respectively the reaction rate. For low-energy collisions involving dihydrogen molecules, a change of the H_2 vibrational state is energetically not accessible. In order to assess the role of the anisotropy of the interparticle interaction, it is useful to expand the angular dependence of the interaction potential into Legendre polynomials,

$$V(R,\theta) = V_0(R) + V_2(R)\cos^2\theta + \cdots .$$

Here, R and θ denote the interparticle separation and the angle between intermolecular axis and collision axis, respectively. A comparison of the magnitude of $V_2(R)$ to the rotational constant of the molecule yields, for H_2, energy equivalents of 1 K versus roughly 200 K. The collision is thus highly unlikely to induce changes of the rotational state. Nevertheless, the anisotropy of the interaction does affect the reaction rate! It determines the occurrence of quantum scattering resonances in slow barrier-less reactions [3] and dictates the scaling of the reaction rate coefficient with collision energy in fast barrier-less reactions [16].

The role of the anisotropy for the occurence of quantum scattering resonances is most readily understood using an adiabatic separation of interparticle motion and internal molecular rotation [17]. Denoting the eigenstates of the internal molecular rotation by $|j, m_j\rangle$, the potential energy operator is effectively diagonal in this basis (reflecting the absence of rotational state changing collisions), but the diagonal matrix elements differ for $j = 0$ and $j = 1$: While the matrix element for $j = 0$ contains only the isotropic part of the interaction, $V_0(R)$, those for $j = 1$ also depend on the anisotropy, $V_2(R)$. This allows one to directly probe the role of the anisotropy

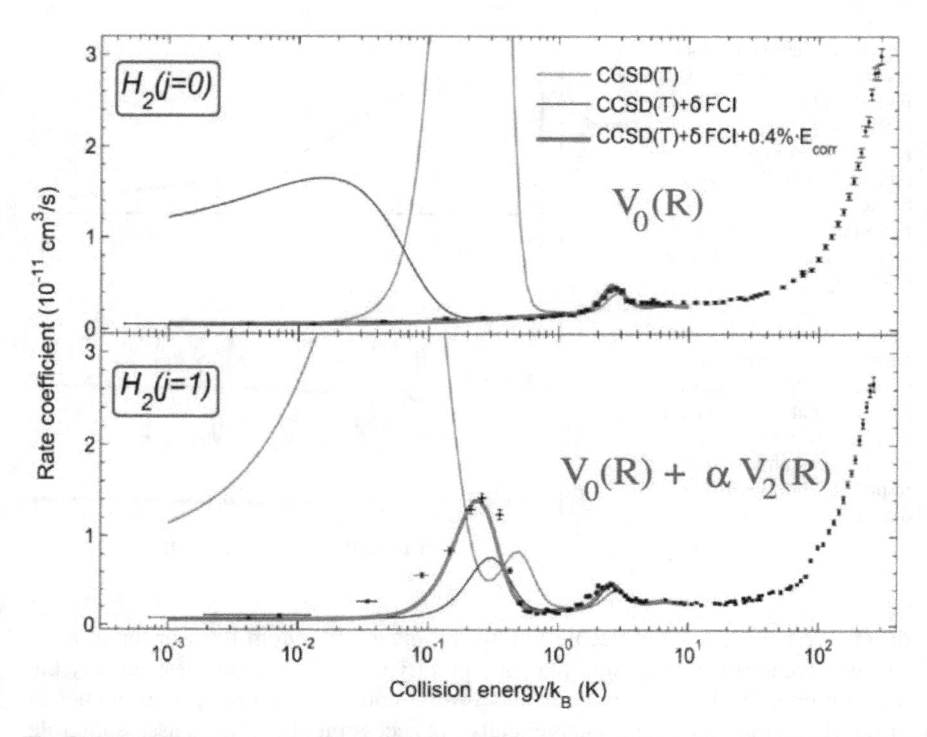

Fig. 2 Rate coefficient for Penning ionization of dihydrogen colliding with metastable helium: The tunneling resonance at a collision energy of $k_B \times 270\,\text{mK}$ appears only if the molecule is rotationally excited. This is due to the anisotropy of the interparticle interaction (equal to V_2 in leading order). Adapted from Ref. [3]

in atom-molecule collisions via quantum scattering resonances [3], provided the molecule can be selectively prepared in $j = 0$ or $j = 1$.

Rotational state selection in H_2 molecules is made possible by the large rotational splitting together with the nuclear spin symmetry. The latter implies that para-H_2 must have an even rotational wavefunction, thus $j = 0$, whereas the rotational wavefunctions for ortho-H_2 are odd, i.e., $j = 1$ for ortho-dihydrogen. Higher rotational levels are not populated in molecular beam experiments due to the large rotational splitting. Samples of almost pure para-H_2 are readily obtained by catalytic ortho-para conversion. It is thus comparatively straightforward to carry out molecular beam studies with rotational state selection [3, 16].

Repeating the Penning ionization experiments of Ref. [13] with dihydrogen molecules in a well defined rotational state led to the surprising observation, cf. Fig. 2, of the low-energy resonance (at a collision energy of $k_B \times 270\,\text{mK}$) occuring only for ortho-H_2, i.e., only for rotationally excited molecules. A theoretical model capturing this phenomenon requires spectroscopic accuracy. Systematic corrections to high-level state-of-the-art quantum chemistry combined with coupled channels calculations allow for reaching this accuracy. This is illustrated by the green, blue and red curves in Fig. 2, depicting the results obtained with a potential energy surface derived by coupled cluster theory with single, double and non-iterative triple excitations (green), including the full CI correction (blue) and further improved by uniformly scaling the correlation energy by 0.4% (red) [3].

The excellent agreement between the theoretical results and the experimental data observed in Fig. 2 allows for a more indepth examination of the role of the anisotropy. In the theoretical model, we can easily modify the relative weight of the anisotropic part of the interaction potential, denoted by α in Fig. 2. Increasing α up to 50% does not introduce a noticeable change on the results obtained with para-H_2, i.e., molecules in their rotational ground state [3]. On the other hand, the low-energy resonance observed only for rotationally excited H_2 in Fig. 2 depends very sensitively on the scaling factor of the anisotropic potential, shifting to lower energies with decreasing anisotropy [3].

The reason underlying this behavior can be unveiled using the adiabatic approximation mentioned above [17]. As is often the case with perturbation theory, it does not yield a quantitative description but qualitatively provides the correct picture. When examining the adiabatically separated scattering channels, a bound state just below the dissociation threshold occurs for rotational ground state para-dihydrogen (with $\ell = 3$, $j = 0$, and $J = 3$). For rotationally excited ortho-dihydrogen, the anisotropic part of the interparticle interaction potential introduces an energy shift that is added to the effective potential. This pushes, what is a weakly bound state for para-dihydrogen, above the dissociation threshold for ortho-dihydrogen, turning the bound state into a shape resonance (with $\ell = 3$, $j = 1$, and $J = 3$) [3], thus solving the riddle of resonance (dis)appearance.

To conclude this section, quantum scattering resonances testify to the emergence of the wave nature of matter in "cold" collisions. Using dihydrogen molecules in merged beam studies allows for simple rotational state selection which in turn can be used to probe the anisotropic part of the interparticle interaction governing Penning ionization reactions [3, 16]. While an excellent agreement between theory and experiment can be reached when including appropriate corrections for electron correlations, the calculation of quantum resonances involving metastable states continues to present a significant challenge even for the highest available levels of first principles based theory.

3 Phase-Protection in Fano-Feshbach Resonances

After understanding the dramatic effects that very small shifts in energy may have on a shape resonance, in the present section, another type of quantum resonance will highlight the sensitivity of resonances to small changes in phase. A Fano-Feshbach resonance describes the decay of a bound quantum state due to coupling with a continuum of scattering states. In contrast to shape resonances, it involves two distinct scattering channels characterized by different quantum numbers, as sketched in Fig. 1. In the original theory, the coupling between bound and continuum states described an effective nucleon-nucleon interaction [18], resp. configuration interaction in autoionization [19]. In recent years, Fano-Feshbach resonances due to the hyperfine interaction between different nuclear spin states have attracted attention as key tool for control in ultracold gases [20]. The present example considers rovibrational predissociation resonances due to the spin-orbit interaction [21].

Predissociation resonances may be populated by associative ionization which accompanies a Penning ionization reaction, provided the dependence of the Penning ionization rate on interparticle separation matches the ionic potential energy curve [15]. The ionization then populates bound levels of noble gas diatomic molecules such as $HeAr^+$, $NeAr^+$, or $HeKr^+$ which may or may not be spin-excited. These seemingly simple molecules with very similar electronic structure possess (spin-excited) predissociation resonances with surprisingly different lifetimes.

In order to estimate lifetimes, one typically inspects the coupling responsible for the decay. When comparing predissociation in $HeAr^+$ and $NeAr^+$, the spin-orbit splitting Δ in the two cases is identical. The term dominating the coupling between spin-excited and ground states is radial coupling [4] which scales as Δ/μ, where μ is the reduced mass. The lifetime of the resonances then scales with μ^2. Since, for the two diatoms, the reduced masses differ by a factor of approximately four, the lifetime of $NeAr^+$ would be expected to be larger than that of $HeAr^+$ by about an order of magnitude. This expectation derived from scaling arguments ignores, however, a phase dependence of the lifetimes which may entirely alter the picture, as shown next.

A qualitative understanding of the resonance lifetimes can be obtained in first order perturbation theory, using Fermi's golden rule:

$$\tau_\varphi^{-1} \sim |\langle k_{\mathrm{res}}|\mathbf{W}_{\mathrm{coupl}}|\varphi\rangle|^2 , \tag{1}$$

where $|\varphi\rangle$ denotes the quasi-bound state, $\mathbf{W}_{\mathrm{coupl}}$ the coupling operator, and $|k_{\mathrm{res}}\rangle$ the continuum state with scattering momentum k_{res}, in resonance with the quasi-bound state, cf. Fig. 1. Since the continuum state describes the scattering off a potential, a phase shift δ is associated with the position of the repulsive wall. It is this phase shift that determines the value of the complex overlap in Eq. (1). In particular, there exist combinations of k and δ for which the overlap vanishes such that $\tau_\varphi \to \infty$ despite non-zero coupling! We have termed this phenomenon "phase protection" [4]. In fact, it is straightforward to show that vanishing overlap is equivalent to

$$\arg(\tilde{\varphi}(k)) + \delta = m\pi \quad \text{with } m \in \mathbb{Z} \tag{2}$$

with $\tilde{\varphi}(k)$ the Fourier transform of the quasi-bound state, when neglecting the energy dependence of the phase shift δ and assuming s-wave scattering.

In a real molecule, the resonance lifetime will never strictly go to infinity since eventually other decay mechanisms will become relevant. However, the lifetime can indeed become very large. This is shown in Fig. 3 displaying the potential energy curves, derived from spectroscopic data, respectively coupled-cluster calculations,

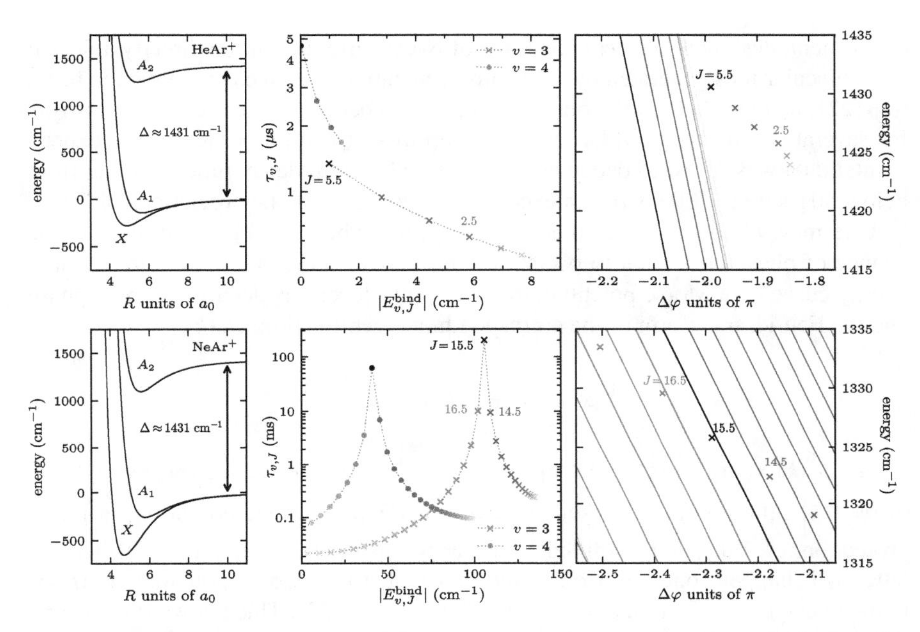

Fig. 3 Potential energy curves (left), lifetimes (center), and resonance positions (right) of predissociation resonances in spin-excited noble gas dimers. Adapted from Ref. [4]

for $HeAr^+$ and $NeAr^+$, together with the lifetimes obtained with a complex absorbing potential and including relativistic and angular couplings [4]. The potential energy curves shown in the left-hand part of Fig. 3 highlight the similarity of the potential energy curves and spin-orbit coupling Δ for the two molecules. In contrast, a striking difference is observed in the middle panel of Fig. 3, where the predissociation lifetimes of spin-excited $NeAr^+$ span more than four orders of magnitude whereas those of $HeAr^+$ differ by only a factor of 10. This difference is rationalized in terms of the condition for phase protection, cf. Eq. (2), visualized in the right part of Fig. 3. The lines indicate the combinations of k (respectively scattering energy) and δ for which the overlap in Eq. (1) vanishes. For $NeAr^+$, several resonance positions are located very close to those lines. The closer the resonance position to its corresponding phase protection line, the larger becomes the corresponding lifetime. In contrast, for $HeAr^+$, none of the resonances fulfills even approximately the condition for phase protection which is why the lifetimes span a much smaller range.

Experimental evidence for the lifetimes of spin-excited $NeAr^+$ spanning several orders of magnitude is provided by two experiments [4], using molecular beams. First, velocity-map imaging (VMI) of the reaction products of Penning and associative ionization of argon by metastable helium revealed predissociation for $HeAr^+$ to occur on a timescale smaller than the time of flight of the VMI setup, of the order of 10 μs. In contrast, the predissociation feature was missing in the VMI for $NeAr^+$, suggesting lifetimes significantly larger than the time of flight [4]. A second experiment, designed to assess the range of $NeAr^+$ lifetimes more directly, injected the molecular ions generated by associative ionization into an electrostatic ion beam trap [22] and recorded oscillations of the molecules between two electrostatic mirrors for several hundreds of milliseconds via trap loss of neutral particles. The experimental data was found to decay non-exponentially, with decay times ranging from below 50 μs up to 100 ms [4], in excellent agreement with the predictions of Fig. 3.

One may wonder whether there is a way to predict, for a given molecule, the chance of phase protection to occur. Assuming one can approximate the potential energy curve by a Morse potential, the lifetime of level v is determined by a phase and the Fourier transform of the corresponding eigenfunction, $\tilde{\psi}_v(k)$,

$$\tau_v^{-1} \propto \sin\left(\delta + \phi_v(k)\right)^2 \left|\tilde{\psi}_v(k)\right|^2 ,$$

where $\phi_v(k) = \arg[\tilde{\psi}_v(k)]$. On top of a smooth variation with k (or energy) due to $\left|\tilde{\psi}_v(k)\right|^2$, the lifetimes τ_v oscillate and vanish whenever the condition for phase protection, cf. Eq. (1), is fulfilled. In order to answer the question how often the latter will happen, one can exploit the fact that, for a Morse potential, $\tilde{\psi}_v(k)$ is known analytically, in terms of complex Γ-functions [23]. This allows for deriving the scaling of the lifetimes with the parameters of the Morse potential and the reduced mass: The number of times τ_v vanishes, and thus the chance for phase protection, increases with μ as well as the well depth and equilibrium distance and decreases with the potential width [4]. This agrees well with the observations for $NeAr^+$ and $HeAr^+$ in Fig. 3 since the reduced mass and the well depth are larger for $NeAr^+$ than for $HeAr^+$, while the position of the minimum and the potential width are very similar. Moreover, this scaling argument predicts an isotope effect on predissociation lifetimes which has indeed been observed experimentally for N_2^+ [24] and Ne_2^+ [25].

To summarize this section, phase protection refers to the fact that the lifetime of a quasi-bound quantum state can become very, very large despite non-zero coupling with a continuum of scattering states. The protection from decay occurs whenever the relative phase in the overlap of bound and scattering state becomes a multiple of π. Experiments with spin-orbit excited noble gas dimers have confirmed our theoretical prediction of phase protection. The probability of phase protection increases with reduced mass, well depth and equilibrium distance of the potential supporting the quasi-bound state, providing a blueprint to identify quantum states that are intrinsically protected against undesired decay.

4 Photoelectron Circular Dichroism and Its Coherent Control

The third example showcases a quantum effect that is observed in experiments with molecular beams when circularly polarized light ionizes molecules which are chiral [7]. Remarkably, when a molecule is chiral, i.e., when its nuclear scaffold exhibits a handedness, ionization of randomly oriented molecules with right circularly polarized light does not yield the same photoelectron angular distribution as that with left circularly polarized light [26]. The difference between the photoelectron angular distributions obtained with left and right circularly polarized light is called photoelectron circular dichroism (PECD). Instead of exchanging the polarization direction of the light, one can also exchange the handedness of the molecule to observe PECD [26]. Unlike other dichroic effects, PECD does not involve a magnetic dipole moment and is obtained merely within the electric dipole approximation of the light-matter interaction [26].

First order perturbation theory for the photoionization cross section reveals the mechanism underlying the dichroic effect [26]: For a chiral molecule, the photoionization matrix elements with opposite spatial orientation do not cancel when averaging over the Euler angles. This gives rise to terms in the cross section that are odd under inversion of the polar angle and thus to dichroism. More intuitively, the photoionization cross section involves two rotations—of the polarization axis into the molecular frame and of the photoelectron momentum into the lab frame—which are sufficient to be sensitive to the handedness of the molecular scaffold. This geometric picture can be made more rigorous by noticing that the angle-resolved photoionization cross section is a vector observable which can be expressed as a triple product [27]. For non-coplanar vectors forming the triple product, the observable becomes enantiosensitive [27].

While theoretically predicted in the mid-1970s [26] and first observed with synchrotron radiation in the early 2000s [28], femtosecond laser pulses driving multiphoton ionization have made PECD accessible in table-top experiments [7, 29–32]. Since multi-photon ionization probes intermediate electronically excited states [7, 33], a theoretical description beyond first order in the perturbation theory for the light-matter interaction is called for. At the same time, the bicyclic ketones with which the experiments have been carried out, for example, fenchone, camphor, or limonene, are amenable to a high-level treatment of their electronically excited states. In contrast, modeling the photoionization continuum from first principles is rather challenging.

A way to address this challenge, applied to the specific example of resonantly enhanced (2+1) multi-photon ionization (REMPI) of fenchone and camphor [5], separates the non-resonant two-photon excitation from the one-photon ionization. The former can be described with coupled cluster theory whereas for the latter a single active electron approach using hydrogenic orbitals captures the essential physics of the photoelectron moving in a Coulombic potential [5]. Neglecting any coherent effects during the excitation, the ionization probes an anisotropic distribution of electronically excited molecules. This is due to the anisotropy of the two-photon absorption tensors of the molecules [5]. The properties of the two-photon absorption alone are, however, not sufficient to determine which intermediate state is probed by the REMPI process. Based on energetic arguments, there are five possible candidate states, only one of which is ruled out when using the information of the two-photon absorption tensors in the calculation of the photoionization cross section [5]. When including also the properties of the probed electronically excited states, only one out of the five candidates is in agreement with the experimental data [5]. Interestingly, the theoretically predicted intermediate state differs for fenchone compared to camphor [5]. This might explain the different photoelectron angular distributions and different PECD observed for these two chemically very similar molecules [7].

There are many intriguing questions that arise in the context of PECD, for example whether the effect is determined by the initial, the final, or the intermediate state of the process. In order to answer such questions, a more rigorous description of PECD and related effects in the ionization of chiral molecules is required. In particular, a better description of the photoionization continuum including electron correlation effects and the ability to treat coupled electronic and vibrational motion are called for. At the same time, PECD provides a very convenient experimental handle to probe the chiroptical response of molecules, and it is natural to ask whether this response can be enhanced by suitably shaping the ionizing field, in the spirit of coherent control of photoinduced dynamics [34, 35].

This question can be answered by combining the time-dependent configuration interaction singles (CIS) method for the electronic structure [36], second order perturbation theory for the light-matter interaction, and parameter optimization for the electric field using sequential parametrization updates [37]. Figure 4 illustrates the ionization pathways that are accounted for in this description. Pathways ending at the same final kinetic energy of the photoelectron (within the spectral bandwidth of the pulse) can interfere with each other, cf. the red and green arrows in Fig. 4. This can be thought of as a spectral realization of a double-slit experiment, with the additional advantage that the relative phase between the two pathways can be adjusted by properly tuning the ionizing electric field in its amplitude and phase [35].

A bichromatic pulse driving the red and green ionization pathways in Fig. 4 is thus a natural starting point for optimizing the electric field. When ionizing a chiral methan derivate with such a pulse, assuming a flat spectral phase, a PECD of about 4% (relative to the isotropic ionization yield) is obtained [6]. When optimizing the

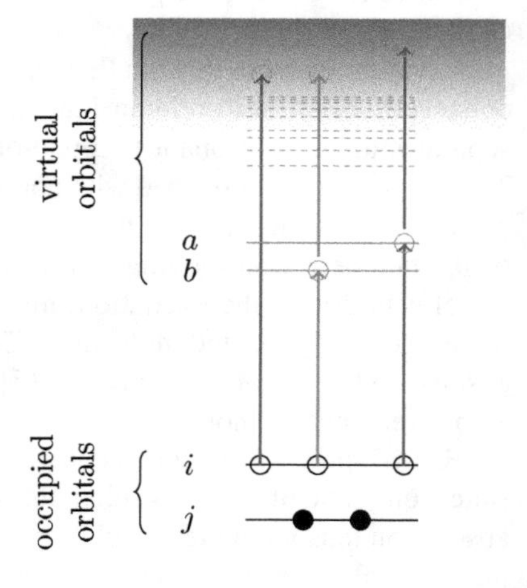

Fig. 4 One-photon and two-photon ionization pathways within the time-dependent CIS framework

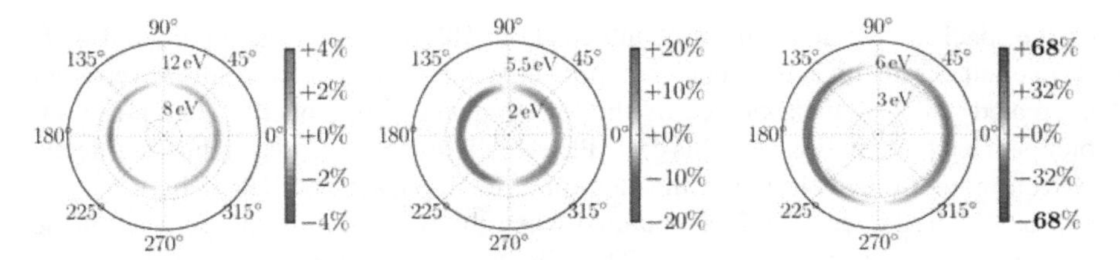

Fig. 5 PECD after ionization of CHBrClF with a bichromatic guess pulse with flat spectral phase (left), optimized bichromatic (center) and freely optimized (right) electric fields. The percentage is taken with respect to the isotropic yield, and the increase is due to interference between one-photon and two-photon pathways (center), respectively between different two-photon pathways probing different intermediate states (right). Adapted from Ref. [6]

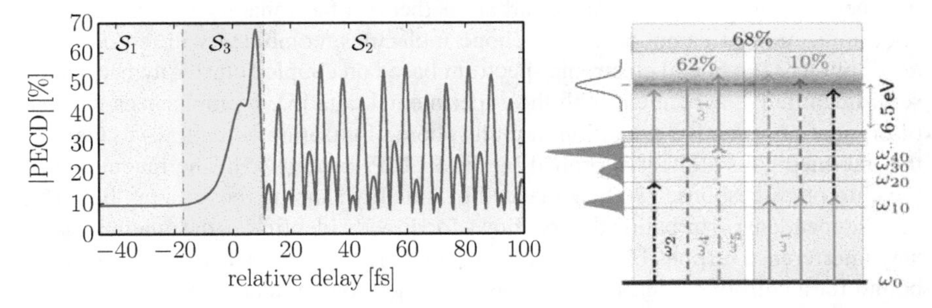

Fig. 6 The freely optimized field drives various two-photon ionization pathways that, depending on the linear chirp of the low-frequency component of the ionizing field, respectively the time-delay between low-frequency and high-frequency pulse compontents, interfere constructively or destructively. Adapted from Ref. [6]

field, constraining the pulse to be bichromatic, the PECD is pushed up to about 20%. This is shown in the center panel of Fig. 5. Compared to the initial guess pulse, the photon energies are lowered and, more importantly, the spectral phases of the two pulse components are adjusted to ensure that one-photon and two-photon ionization pathways interfere constructively. In contrast, with a flat spectral phase, the interference is mainly destructive. Indeed, constructive ionization of one-photon and two-photon pathway makes up for 17% out of the calculated 20% of PECD [6].

An even larger increase in PECD, up to almost 70%, is obtained when freely optimizing the ionizing electric field [6], cf. right panel of Fig. 5. This increase is driven by interference between various two-photon ionization pathways probing different intermediate, electronically excited states, illustrated in Fig. 6 (right). The optimized field turns out to have two spectral components, with the lower one exhibiting a linear spectral phase [6]. Such a "chirp" provides a very convenient handle to analyze the interference pattern in more detail, since changing the chirp rate, i.e., the slope of the spectral phase, amounts to changing the relative time delay between the low-frequency and the high-frequency component of the optimized pulse. If the low-frequency component precedes the high-frequency one, the PECD does not depend on the specific time delay and amounts to about 10%, cf. red-shaded parts in Fig. 6. If, in contrast, the high-frequency component arrives first, an electronic wavepacket is created which is then ionized by the low-frequency component of the pulse, cf. blue-shaded parts in Fig. 6. In this case, PECD sensitively depends on the time-delay, reflecting the time evolution of the electronic wave packet, oscillating between 6% and 62%. The largest PECD is obtained when the two spectral components of the optimized pulse overlap in time [6]. This allows for maximum interference between all the two-photon ionization pathways, highlighted by the yellow-shaded parts in Fig. 6.

In summary, photoelectron circular dichroism is a sensitive chiroptical probe of electron dynamics in a chiral potential, measured in experiments with molecular beams. Time-independent perturbation theory of resonantly enhanced (2+1) photoionization of camphor and fenchone molecules, combined with an *ab initio* description of the bound electronic spectrum based on coupled cluster theory, yields semi-quantitative agreement with the experimental data [5]. It emphasizes the role of orientation-selective excitation in multi-photon ionization and allows to identify the intermediate state that is probed by the REMPI process [5]. The magnitude of the chiroptical response can be enhanced by suitably shaping the ionizing field [6]. Time-dependent perturbation theory allows for directly identifying the quantum pathway interference responsible for the enhancement [6]. Whether there exists an upper bound for a chiroptical response such as photoelectron circular dichroism is one of the many open questions in the quantum control of molecular chirality.

5 Conclusions

This chapter reviewed three examples of quantum effects beyond entanglement — the (dis)appearance of a tunneling resonance in cold collisions, the protection of predissociation resonances by a phase, and the circular dichroism observed in the photoelectron spectrum of chiral molecules. In each case, a theoretical description based on first principles was key to elucidating the rather surprising observations made in experiments with molecular beams. The examples thus testify to the intriguing nature of quantum mechanics as well as to the topicality of Otto Stern's legacy.

Acknowledgements I would like to thank my students and postdocs, in particular Alexander Blech, Daniel Reich, Esteban Goetz and Wojtek Skomorowski, for their dedication and determination in carrying out the work that I have reviewed here. I am indebted to my colleagues Edvardas Narevicius and Robert Berger for many years of inspiring and fruitful collaboration. I am grateful for a Rosi and Max Varon Visiting Professorship to the Weizmann Institute of Science and for financial support from the German-Israeli Foundation (grant no. 1254), the State Hessen Initiative for the Development of Scientific and Economic Excellence (LOEWE) within the focus project Electron Dynamic of Chiral Systems (ELCH), and the Deutsche Forschungsgemeinschaft (Projektnummer 328961117—SFB 1319). This work was carried out while I was at the University of Kassel; and I would like to thank my Kassel colleagues for creating a friendly and stimulating research atmosphere.

References

1. J.P. Dowling, G.J. Milburn, Philos. Trans. R. Soc. A **361**, 1655 (2003)
2. A. Acín et al., New J. Phys. **20**, 080201 (2018)
3. A. Klein et al., Nat. Phys. **13**, 35 (2016)
4. A. Blech et al., Nat. Commun. **11**, 999 (2020)
5. R.E. Goetz, T.A. Isaev, B. Nikoobakht, R. Berger, C.P. Koch, J. Chem. Phys. **146**, 024306 (2017)
6. R.E. Goetz, C.P. Koch, L. Greenman, Phys. Rev. Lett. **122**, 013204 (2019)
7. C. Lux, M. Wollenhaupt, T. Bolze, Q. Liang, J. Köhler, C. Sarpe, T. Baumert, Angew. Chem. Int. Ed. **51**, 5001 (2012)
8. O. Dulieu, A. Osterwalder (eds.), *Cold Chemistry* (The Royal Society of Chemistry, Theoretical and Computational Chemistry Series, 2018)
9. R.V. Krems, *Molecules in Electromagnetic Fields* (Wiley, Hoboken, NJ, 2019)
10. H.R. Sadeghpour, J.L. Bohn, M.J. Cavagnero, B.D. Esry, I.I. Fabrikant, J.H. Macek, A.R.P. Rau, J. Phys. B **33**, R93 (2000)
11. Y. Shagam, E. Narevicius, J. Phys. Chem. C **117**, 22454 (2013)
12. A. Osterwalder, E.P.J. Techn, Instrum. **2**, 10 (2015)
13. A.B. Henson, S. Gersten, Y. Shagam, J. Narevicius, E. Narevicius, Science **338**, 234 (2012)
14. M. Costes, C. Naulin, Chem. Sci. **7**, 2462 (2016)
15. P.E. Siska, Rev. Mod. Phys. **65**, 337 (1993)
16. Y. Shagam, A. Klein, W. Skomorowski, R. Yun, V. Averbukh, C.P. Koch, E. Narevicius, Nat. Chem. **7**, 921 (2015)
17. M. Pawlak, Y. Shagam, E. Narevicius, N. Moiseyev, J. Chem. Phys. **143**, 074114 (2015)
18. H. Feshbach, Ann. Phys. **5**, 357 (1958)
19. U. Fano, Phys. Rev. **124**, 1866 (1961)
20. C. Chin, R. Grimm, P. Julienne, E. Tiesinga, Rev. Mod. Phys. **82**, 1225 (2010)
21. A. Carrington, T.P. Softley, Chem. Phys. **92**, 199 (1985)
22. D. Zajfman, O. Heber, L. Vejby-Christensen, I. Ben-Itzhak, M. Rappaport, R. Fishman, M. Dahan, Phys. Rev. A **55**, R1577 (1997)
23. M. Bancewicz, J. Phys. A **31**, 3461 (1998)
24. T.R. Govers, C.A. van de Runstraat, F.J. de Heer, J. Phys. B **6**, L73 (1973)
25. K. Gluch, J. Fedor, R. Parajuli, O. Echt, S. Matt-Leubner, P. Scheier, T.D. Märk, Eur. Phys. J. D **43**, 77 (2007)
26. B. Ritchie, Phys. Rev. A **13**, 1411 (1976)
27. A.F. Ordonez, O. Smirnova, Phys. Rev. A **98**, 063428 (2018)
28. N. Böwering, T. Lischke, B. Schmidtke, N. Müller, T. Khalil, U. Heinzmann, Phys. Rev. Lett. **86**, 1187 (2001)
29. C.S. Lehmann, N.B. Ram, I. Powis, M.H.M. Janssen, J. Chem. Phys. **139** (2013)
30. C. Lux, M. Wollenhaupt, C. Sarpe, T. Baumert, ChemPhysChem **16**, 115 (2015)
31. R. Cireasa et al., Nat. Phys. **11**, 654-658 (2015)
32. A. Comby et al., J. Phys. Chem. Lett. **7**, 4514 (2016)
33. A. Kastner, T. Ring, B. C. Krüger, G.B. Park, T. Schäfer, A. Senftleben, T. Baumert, J. Chem. Phys. **147**, 013926 (2017)
34. S.A. Rice, M. Zhao, *Optical Control of Molecular Dynamics* (Wiley, New York, 2000)
35. M. Shapiro, P. Brumer, *Quantum Control of Molecular Processes*, 2nd, revised and enlarged edition edition (Wiley Interscience, New York, 2012)
36. L. Greenman, P.J. Ho, S. Pabst, E. Kamarchik, D.A. Mazziotti, R. Santra, Phys. Rev. A **82**, 023406 (2010)
37. R.E. Goetz, M. Merkel, A. Karamatskou, R. Santra, C.P. Koch, Phys. Rev. A **94**, 023420 (2016)

9

Precision Physics in Penning Traps using the Continuous Stern-Gerlach Effect

Klaus Blaum and Günter Werth

1 Introduction

"A single atomic particle forever floating at rest in free space" (H. Dehmelt) would be the ideal object for precision measurements of atomic properties and for tests of fundamental theories. Such an ideal, of course, can ultimately never be achieved. A very close approximation to this ideal is made possible by ion traps, where electromagnetic forces are used to confine charged particles under well-controlled conditions for practically unlimited time. Concurrently, sensitive detection methods have been developed to allow observation of single stored ions. Various cooling methods can be employed to bring the trapped ion nearly to rest. Among different realisations of ion traps we consider in this chapter the so-called Penning traps which use static electric and magnetic fields for ion confinement. After a brief discussion of Penning-trap properties, we consider various experiments including the application of the "continuous Stern-Gerlach effect", which have led recently to precise determinations of the masses and magnetic moments of particles and antiparticles. These serve as input for testing fundamental theories and symmetries.

2 Penning-Trap Properties

The Penning traps used in the experiments described herein consist of a symmetric stack of cylindrical electrodes as shown in Fig. 1.

K. Blaum
Max-Planck-Institut für Kernphysik, Heidelberg, Germany

G. Werth (✉)
Johannes-Gutenberg-Universität, Mainz, Germany
e-mail: werth@uni-mainz.de

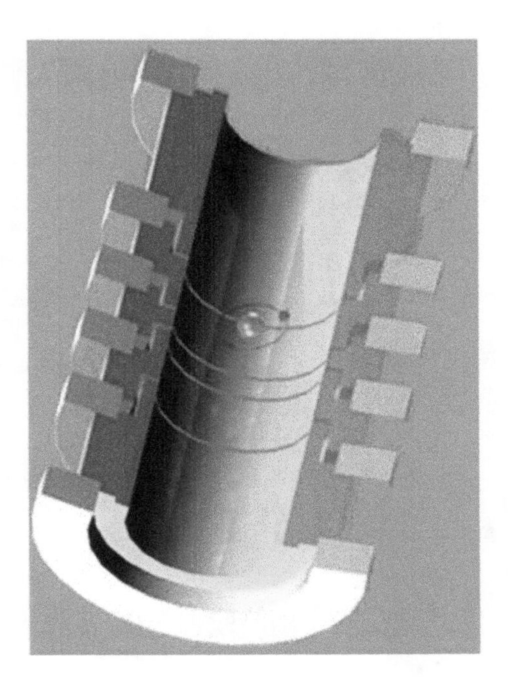

Fig. 1 Schematic of our cylindrical Penning trap holding a single highly charged ion stored in this electrode configuration

A static voltage is applied between the connected outer electrodes and the central ring. At a positive polarity at the endcaps a positively charged particle is confined in the axial direction. Escape in the radial direction is precluded by a homogeneous magnetic field directed along the trap's axis. Additional voltages applied to the so-called correction electrodes, placed between the ring and endcaps, serve to provide an electric potential which depends—at least for small amplitudes of the trapped particle—only on the square of the distance from the trap's centre. This is a prerequisite for achieving high resolution in the determination of the motional frequencies of the trapped ions. It is further required to avoid perturbation of the confined ion by collisions with background gas molecules. In our case, the trap electrodes and their container box are in thermal contact with a liquid helium bath. The cryopumping results in a residual pressure of less than 10^{-16} mbar. As a result, not even a single collision of the trapped ion with a neutral molecule was observed during typical trapping times of several months.

The ion's motion in the trap arrangement described above can be calculated analytically. As a result one obtains a superposition of three harmonic oscillations: An axial one at frequency ν_z, which depends on the ions mass M, the size of the trap, and the voltage, and two radial oscillations with frequencies ν_+ and ν_-. The frequency ν_+ is near the cyclotron frequency $\nu_c = qB/(2\pi M)$ of the free ion with charge q in the magnetic field B, slightly perturbed by the presence of the electric trapping field. The centre of this motion orbits around the trap centre at a low "magnetron frequency" ν_-. Figure 2 shows a sketch of the ion's motion. An important relation connects the motional frequencies to the free ion's cyclotron frequency [1]:

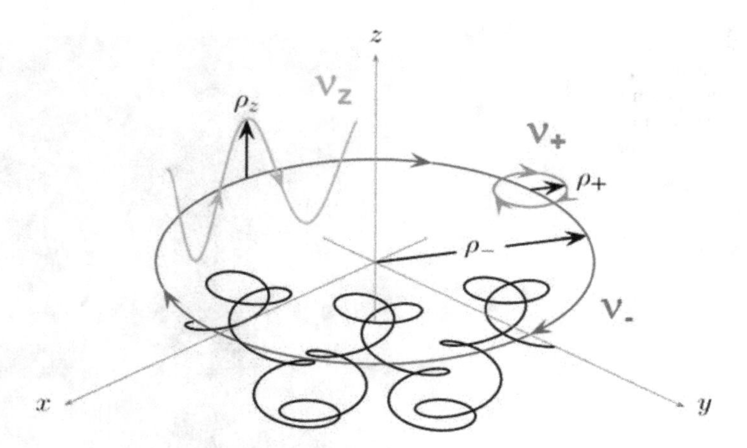

Fig. 2 Sketch of the ion's motion in a Penning trap. For details see text

$$\nu_+^2 + \nu_-^2 + \nu_z^2 = \nu_c^2. \tag{1}$$

Further details on Penning traps can be found in Refs. [2, 3].

3 Single Ion Detection by Induced Image Currents

Achieving high precision in trap experiments requires the use of single trapped particles to avoid perturbations by Coulomb interaction with other ions. The standard way to detect single trapped particles is by observation of laser induced fluorescence. This requires, however, optical transitions in the ions which are in reach of laser wavelengths. This is not the case for highly charged ions or elementary particles such as electrons or protons and their antiparticles. In these cases, detection can be performed by the image current that the oscillating ion induces in the trap's electrodes [4]. This current is on the order of a few fA and requires very sensitive detection methods. This can be realized by a superconducting high-Q tank circuit, kept at the temperature of the surrounding He-bath and tuned to the resonance frequency of the ion oscillation. Figure 3 shows a scheme of the detection of a radial frequency. The noise power of the circuit can be amplified and Fourier analysed. In case of

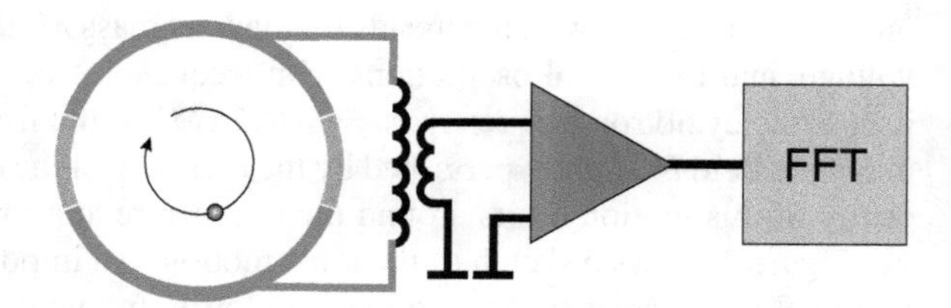

Fig. 3 Scheme for detection of the radial ion oscillation by a superconducting tank circuit attached to segments of the trap's split ring electrode

an ion present in the trap, the Fourier spectrum shows a maximum at the resonance frequency of the ion oscillation, as shown in Fig. 4.

The ion signal as shown in Fig. 4 indicates that the ion's kinetic energy is well above the thermal energy of the circuit. In order to reduce the ion's energy as required for high-precision measurements, the ion is kept in resonance with the circuit. The extra energy which the hot ion transfers into the circuit is then dissipated into the helium bath (resistive cooling) [5]. As consequence the ion adopts the temperature of the environment. The signal is then converted into a minimum in the noise spectrum as shown in Fig. 5. This can be understood based on the fact that the equivalent electronic circuit of the oscillating ion is a series resonance circuit which shortcuts the noise power of the detection circuit at its resonance frequency.

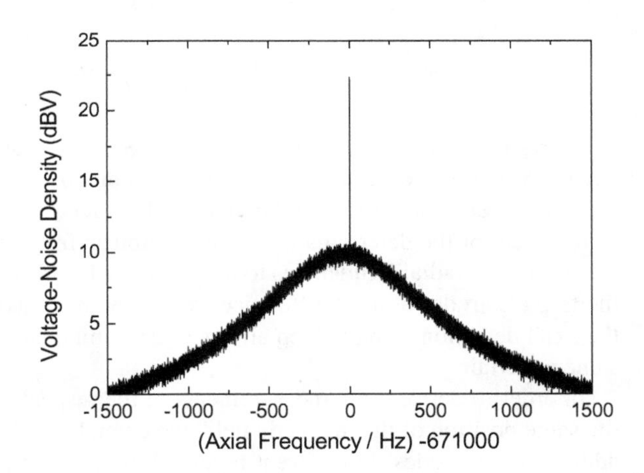

Fig. 4 Fourier analysis of the axial detection resonance circuit showing the induced image current of a single trapped ion on top of the Johnson noise of the detection circuit

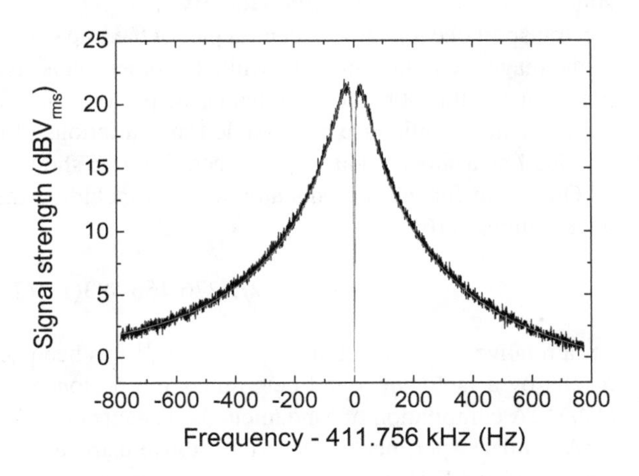

Fig. 5 Signal of a single ion's axial resonance in thermal equilibrium with the detection system immersed in a liquid He bath

4 The Masses of the Proton and Antiproton

The ion signals such as those shown in Fig. 5 can be used for high-precision mass measurements. As recent examples, we consider the atomic masses of the proton and the antiproton. Their comparison serves as a test of the CPT invariance theorem. Quite general determination of atomic masses requires the comparison with the standard atomic mass, the carbon atom. The comparison is performed through measurements of the cyclotron frequencies of the particle under investigation $\nu_c(P) = eB/(2\pi M_P)$ (e.g. the proton P with mass M_P) and that of a carbon ion of charge state q: $\nu_c(C^{q+}) = qB/(2\pi M_C^{q+})$ in the same magnetic field B provided by a superconducting solenoid. In the case of the proton we used C^{6+} and correct the measured frequency by the masses of the missing electrons and their respective binding energies:

$$M_P = \frac{e}{q}\frac{\nu_C(C)}{\nu_C(P)} M_C \tag{2}$$

By resolving the central part of the ion detection signal as shown in Fig. 5, we can determine the centre frequency with an uncertainty of a few mHz. In our experiments we measure exclusively the axial frequency. In order to determine the radial frequency as required for the determination of the cyclotron frequencies according to Eq. (1), we couple the radial frequencies to the axial one by an additional r.f. field applied to the trap electrodes at their difference frequency to the axial. This leads to a split of the axial detection signal which allows to determine the radial frequencies with the same uncertainty.

In order to perform the measurements of the respective cyclotron frequencies at the same position of the magnetic field, we extend our Penning trap by a number of additional electrodes. In different potential minima we can store simultaneously a single proton and a single carbon ion. By changing the potentials at the electrodes we can transport one ion into the central part of the trap structure where the homogeneity of the magnetic field is highest, while the other ion is stored in one of the remaining potential minima. Frequent exchange of protons and carbon ions eliminates to a large extent the influence of possible time variations of the magnetic field which is provided by a superconducting magnet. Figure 6 shows the complete setup.

Our result for the proton's atomic mass including the statistical and systematic uncertainties is [6]

$$M_P = 1.007\ 276\ 466\ 583(15)(29)\ \text{u}. \tag{3}$$

i.e. a relative mass uncertainty of 3×10^{-11} has been achieved. It improves earlier results by a factor of 3 and determines the proton mass value in the most recent CODATA compilation of fundamental constants [7] (Fig. 7).

A similar experiment as described above using a nearly identical setup has been performed at CERN/Geneva, where single antiprotons have been confined and their cyclotron frequency measured [8]. The main difference was that, for comparison with a reference mass, not a carbon ion could be used since it would require a change

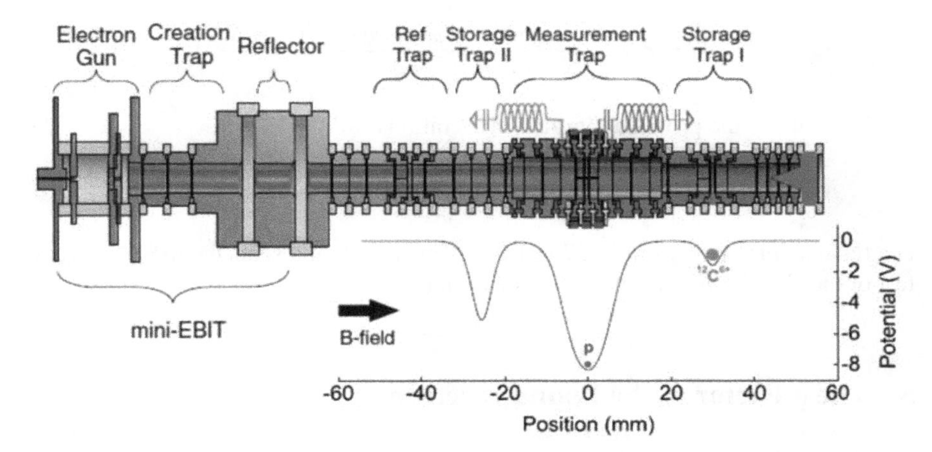

Fig. 6 Setup of the trap configuration for the determination of proton's atomic mass. The left part serves for the creation of protons and carbon ions by electron bombardment of a target. After ion creation and removal of unwanted species, single ions are transferred to one of the potential minima. Measurements are performed in the so-called measurement trap located at the most homogeneous part of the magnetic field. Shown are the resonance circuits attached to the measurement trap's electrode tuned to the different axial frequencies for the proton and C^{6+}. The trap configuration is placed in a hermetically sealed container in thermal contact with a liquid-He bath. The low temperature of the container walls provides a vacuum below 10^{-16} mbar by cryofreezing. Collisions of the stored particles with background molecules are absent for the period of several months which allows a nearly infinitely long perturbation-free storage time

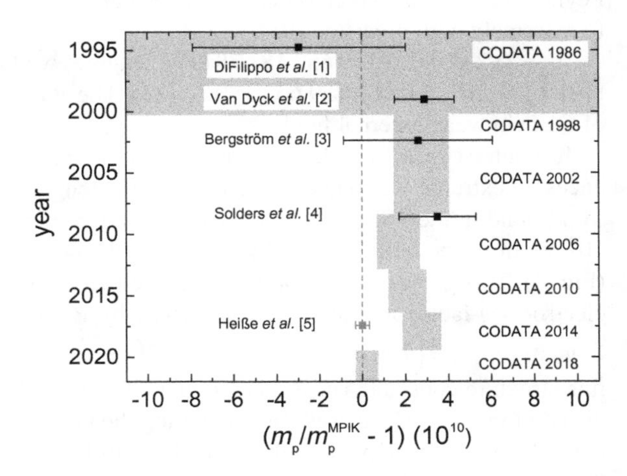

Fig. 7 History of proton mass determinations (*courtesy* F. Heiße)

of the voltage sign at the trap because of the different charge signs of the particles. This, in turn, could lead to uncontrollable errors. Instead, the CERN experiment used the negative hydrogen ion H^-. Taking into account the masses of the 2 additional electrons, their binding energies E_a and E_b as well as the polarizability α_{pol,H^-} of H^-, the antiproton's m_{ap} mass can be compared to the protons mass through

$$m_{(H^-)} = m_{ap}\left(1 + 2\frac{m_e}{m_p} + \frac{\alpha_{pol,H^-B_0^2}}{m_p} - \frac{E_b}{m_p} - \frac{E_a}{m_p}\right) \tag{4}$$

The result of the proton/antiproton mass ratio is

$$(q/m)_{\rm ap}/(q/m)_{\rm p} = -1.000\,000\,000\,001\ (69) \tag{5}$$

at a relative uncertainty level of 7×10^{-11}. This result represents the most stringent test of the CPT invariance in the baryon sector.

5 The *g*-Factor of the Bound Electron

The *g*-factor of the electron is a dimensionless constant which relates the electron's magnetic moment μ_s to the spin S and the Bohr magnetron μ_B:

$$\mu_s = g\mu_B S \tag{6}$$

Dirac's relativistic treatment of the free electron predicts $g = 2$. Experimentally a deviation from 2 is found, which among others gave rise to the theory of Quantum Electrodynamics (QED), which describes the interaction of charged particles with electromagnetic fields by the exchange of virtual photons. Evaluation of Feynman diagrams to high orders allows calculating the *g*-factor of the free electron to extremely high precision [9]:

(g − 2)/2 = 0.001 159 652 181 78 (77). It agrees well with the experimental value [10]: (g − 2)/2 = 0.001 159 652 180 73 (28). The agreement represents the best test of QED in weak external fields.

In contrast to a free electron, an electron bound to an atomic nucleus experiences an extremely strong electric field. The strength of the field for a single electron bound in the 1S-state of a hydrogen-like ion of nuclear charge Z ranges from 10^9 V/cm in the helium ion (Z = 2) to > 10^{15} V/cm in H-like uranium (Z = 92) (Fig. 8). This gives rise to a variety of new effects. The largest change of the bound electron's *g*-factor was analytically derived by Breit (1928) from the Dirac equation: $g_{Breit} = \frac{2}{3}\left(1 + 2\sqrt{1-(Z\alpha)^2}\right) \approx 2(1 - \frac{1}{3}(Z\alpha)^2)$ [11], with $\alpha \approx 1/137$ the fine structure constant. The extremely high electric fields within atoms also require different methods to be used for calculating the QED contributions to the electron's magnetic moment. Feynman diagrams have to be calculated using the solution of the Dirac equation as an electron propagator. Contributions of high orders in $(Z\alpha)$ have been calculated by several authors [12]. In addition, nuclear structure and recoil effects must be considered [13]. Figure 9 summarises the different contributions to the electron's *g*-factor in the ground state of H-like ions as function of the nuclear charge Z.

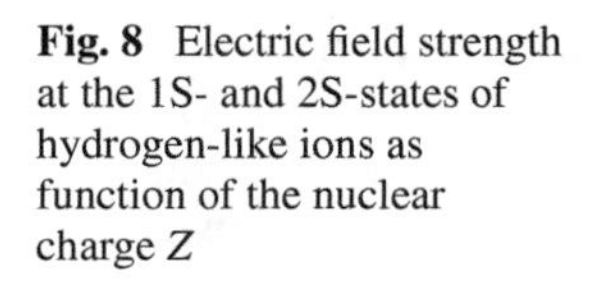
Fig. 8 Electric field strength at the 1S- and 2S-states of hydrogen-like ions as function of the nuclear charge Z

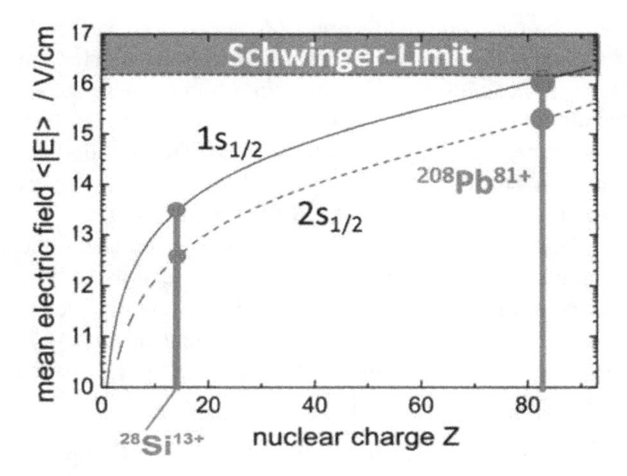

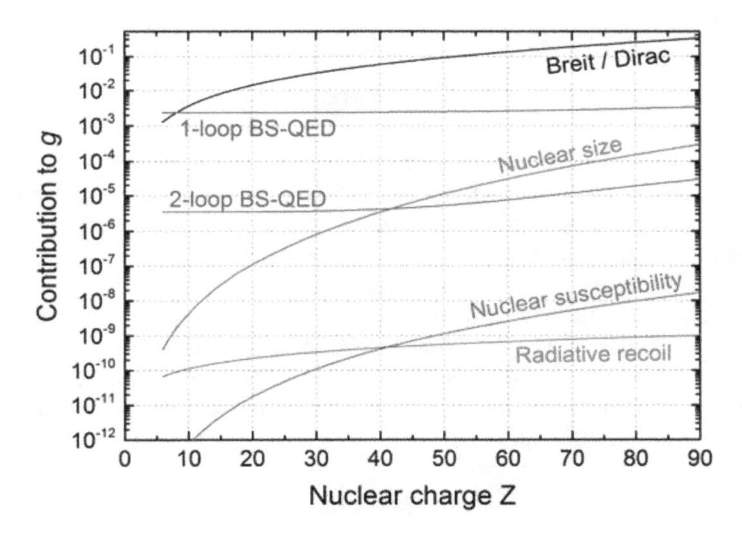

Fig. 9 Contributions from bound-state QED, nuclear size, structure, and recoil to the g-factor of the electron bound in hydrogen-like ions as function of the nuclear charge Z (*courtesy* Z. Harmann)

6 The Continuous Stern-Gerlach Effect

The determination of the electron's g-factor requires the measurement of the spin precession (Larmor) frequency ω_L:

$$\omega_L = \frac{g}{2}\frac{e}{m_e}B \tag{7}$$

The magnetic field strength B can be derived from the measurement of the three motional frequencies as described in Sect. 4. The spin precession, however, does not influence the ion's motion in a homogeneous magnetic field and consequently cannot be detected by observation of the axial ion resonance as only observable in our experimental set-up (see Sect. 3). The required coupling of the spin motion to the ion's oscillation can be provided by an inhomogeneity of the magnetic field. The force of $F = \mu_S gradB$ of a B-field inhomogeneity $gradB$ on the magnetic moment μ_S associated with the spin increases or reduces the electric trapping force of the Penning trap depending of the spin's direction. Consequently, a change in the spin direction leads to a change in the ion's axial frequency. This method has been first employed by Dehmelt in his experiment on the g-factor of the free electron and termed "continuous Stern-Gerlach effect" [14]. It was later adapted to experiments on highly charged ions [15].

The size of the change in the axial frequency upon a change in the spin direction is given by

$$\Delta\omega_z = \frac{g\mu_B B_2}{M\omega_z} \tag{8}$$

where M is the ions mass and B_2 is the quadratic part of the series expansion of the magnetic field $B_Z = B_O + B_2\, z^2 + \ldots$. The magnetic field inhomogeneity in our experiments is produced by a ferromagnetic central ring electrode (nickel) in the Penning trap assembly. It produces a bottle shaped B-field. Odd terms in the series expansion vanish. The measured value of B_2 in our trap is 10 mT/mm^2. The calculated change in the axial frequency, e.g., for H-like $^{28}Si^{13+}$ is 240 mHz in a total oscillation frequency of 412 kHz. The detection of such small frequency changes requires high stability of the electric trapping field. Figure 10 shows that changes in the spin direction, induced by a microwave field in the trap, can be unambiguously detected with nearly 100% probability.

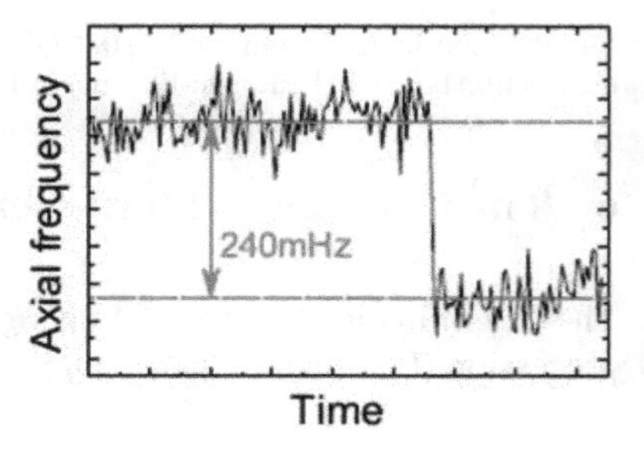

Fig. 10 Change in the axial frequency of a single trapped $^{28}Si^{13+}$ when the spin direction is flipped by microwave-induced transitions. For details see text

7 Measurement of *g*-Factors

Our first application of the continuous Stern-Gerlach effect was the determination of the *g*-factor of the electron bound in hydrogen-like ions. We monitor induced spin flips as shown in Fig. 10 while varying the microwave field's frequency. The maximum spin flip probability occurs when the microwave frequency ω coincides with the Larmor precession frequency ω_L. The *g*-factor can be derived from Eq. (5) when the magnetic field *B* is known. *B* is obtained using the single ion detection signals at the ion's oscillation frequencies as described in Sects. 3 and 4. The *g*-factor then follows from the measurement of ω_L and ω_c:

$$g = 2\frac{\omega_L}{\omega_c}\frac{q_{ion}}{m_{ion}}\frac{m_e}{e} = 2\Gamma\frac{q_{ion}}{m_{ion}}\frac{m_e}{e} \tag{9}$$

Γ is the ratio of the applied microwave frequency ω to the measured cyclotron frequency. However, in order to obtain high precision in the *g*-factor determination we are faced with conflicting requirements: The continuous Stern-Gerlach effect requires a strong inhomogeneity of the *B*-field at the ions position in order to detect induced spin flips with high probability. High accuracy for *B*-field determination on the other side requires a very homogeneous field at the ion's position. In order to resolve this conflict we used a Penning-trap configuration similar to the one shown in Fig. 6 but modified by introducing a nickel ring electrode in one of the storage traps that provides the required *B*-field inhomogeneity. Step 1 of the measurement procedure is the determination of the ion's spin direction in the inhomogeneous trap by introduction of a spin flip and observation of the change in the axial oscillation frequency as illustrated in Fig. 10. With a known spin direction, the ion is then transported into the measurement trap where the oscillation frequency ω_c is determined. Simultaneously the ion is irradiated by a microwave field attempting to change the spin direction. Then the ion is transported back into the inhomogeneous trap and, as in step 1, its spin direction is determined again. A successfully induced spin flip in the measurement trap is monitored in the inhomogeneous trap by the corresponding change in the axial frequency. Frequent repetition of this procedure with varying microwave frequencies and monitoring the spin flip probability for different frequencies results in a resonance curve with a maximum at the Larmor precession frequency ω_L. The fact that ω_L and ω_c are measured at the same position as well at the same time eliminates to a large extent the uncertainties due to time fluctuations of the *B*-field. Figure 11 shows an example of measurements on hydrogen-like $^{28}Si^{13+}$, where the spin flip probability is plotted against the ratio Γ of ω and $\omega_{c.}$. The maximum is at $\Gamma = \omega_L/\omega_{C.}$.

The result of the experiment on $^{28}Si^{13+}$ for the *g*-factor of the single bound electron is $g_{exp} = 1.995\ 348\ 958\ 7\ (5)(3)(8)$ [16]. The numbers in parenthesis correspond respectively to the systematic and statistical uncertainties and the error of the electron mass taken from the CODATA 2012 tables of fundamental constants [17]. The result agrees well with the theoretical value $g_{theo} = 1.995\,348\,958\,0\,(17)$ [18] and represents the most stringent test of QED calculations in strong external fields.

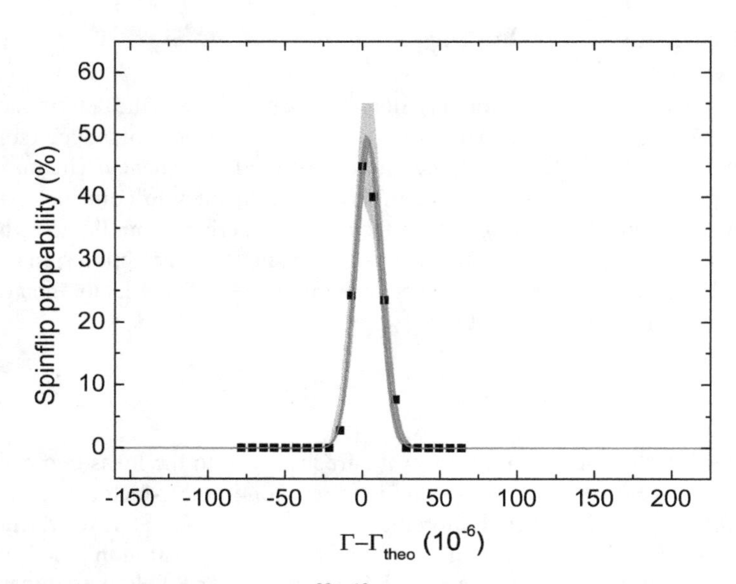

Fig. 11 Measured spin flip probability of $^{28}Si^{13+}$ as a function of the frequency ratio $\Gamma = \omega/\omega_C$. The maximum occurs at $\Gamma = \omega_L/\omega_C$. Γ_{theo} is the theoretical predicted value. Shifts caused by systematic effects are not yet corrected for

Similar results with comparable precision have been obtained on H-like $^{12}C^{5+}$ [19] and $^{16}O^{7+}$ [20]. Also experiments using lithium-like ions $^{40}Ca^{17+}$ and $^{48}Ca^{17+}$ [21], and $^{28}Si^{11+}$ [22] have been performed. Here the interaction of the additional 2 bound electrons modify slightly the value of the *g*-factor. The comparisons between theory and experiment test calculations of the inter-electronic interaction. Most recently the *g*-factor of the boron-like ion $^{40}Ar^{13+}$ has been determined with high precision [23]. The experimental result distinguishes between the conflicting predictions of the contribution of the electron-electron interaction.

Analogously to the electronic *g*-factors, the magnetic moments of the proton and antiproton were obtained. The particular challenge in these experiments had to do with the fact that the magnetic moment of these particles is about 500 times smaller than that of the electron. According to Eq. (6), the corresponding change in the axial frequency is significantly smaller than in the case of hydrogen-like ions. Its observation requires extremely high stability in the trap parameters. A measurable frequency change upon a spin flip was obtained by making the trap diameter about 3 times smaller than in the previous experiments and replacing the Ni-ring by a CoFe-ring and thus obtaining a larger magnetic field inhomogeneity. The result for the proton's and antiproton's magnetic moments are $\mu_p = 2.792\,847\,344\,62(82)\ \mu_N$ [24] and $\mu_{antip} = -2.792\,847\,344\,1(42)\ \mu_N$ [25], with μ_N the nuclear magnetron. The agreement of the two values within the error bars represents a test of the CPT invariance theorem.

8 The Electron Mass

As evident from the result of the g-factor determination in $^{28}Si^{13+}$, the largest contribution in the error budget arises from the uncertainty in the electron mass taken from the CODATA 2012 compilation of fundamental constants. The QED contributions to the electronic g-factor of H-like ions scale approximately with the square of the nuclear charge Z. Since in the case of $^{28}Si^{13+}$ with $Z = 14$ we find an agreement between theory and experiment on the level of 10^{-10} it can be reasonably assumed that in the case of H-like $^{12}C^{5+}$ with $Z = 6$ the QED contributions to the g-factor have been calculated correctly. We can therefore rewrite Eq. (9) between the g-factor and the electron mass as

$$m_e = \frac{g}{2}\frac{e}{q}\frac{v_{\text{cyc}}}{v_{\text{L}}} m_{\text{ion}} \equiv \frac{g}{2}\frac{e}{q}\frac{1}{\Gamma} m_{\text{ion}} \tag{10}$$

We take now the g-factor from theory and determine the electron mass. $^{12}C^{5+}$ is the natural choice as ion since there is virtually no uncertainty in its mass. In our experiment [26] we obtained as new value for the electron mass $m_e = 0.000\,548\,579\,909\,067\,(14)(9)(2)$ a.u. The first two errors are the statistical and systematic uncertainties of the measurement, respectively, and the third one represents the uncertainties of the theoretical prediction of the g-factor and the electron binding energies in the carbon ion. The new value surpasses that of the CODATA 2012 literature value by a factor of 13 and represents the basis for the most recent adjustment of fundamental constants [7].

9 What Comes Next?

At the Max-Planck-Institut für Kernphysik in Heidelberg an electron beam ion trap (EBIT) [27] will allow production of hydrogen-like ions of high nuclear charge Z up to $^{208}Pb^{81+}$. They can be extracted from the EBIT and injected into an improved Penning-trap arrangement (ALPHATRAP) placed at the center of a superconducting magnet [28]. Here Larmor-to-cyclotron frequency ratio measurements can be performed at the $dg/g \approx 10^{-12}$ level of accuracy. This will provide more stringent tests of bound state QED contributions to the electron's g-factor in an extremely strong electric field as evident from Figs. 8 and 9. In addition a new determination of the fine structure constant α by comparison of theoretical and experimental results seems possible. Figure 12 shows a sketch of the setup.

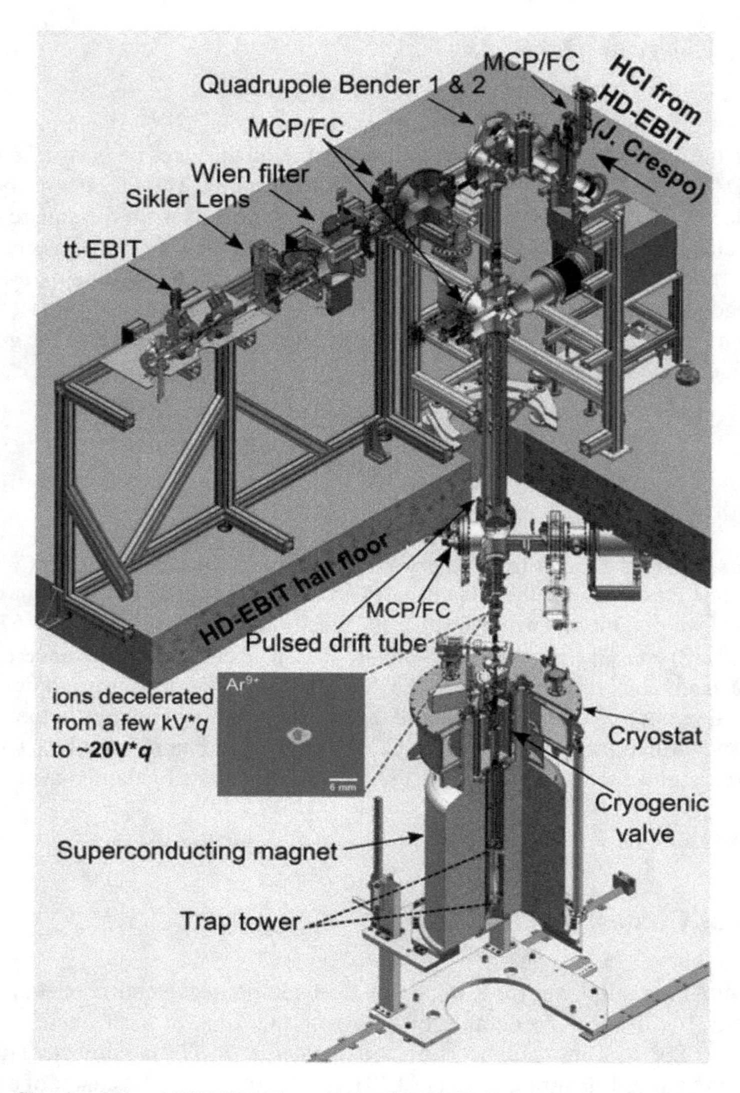

Fig. 12 Sketch of the ALPHATRAP set-up at MPIK Heidelberg to produce H-like ions of high nuclear charge from an EBIT to be injected into a Penning trap arrangement for g-factor determination [28]. A first experiment on ion $^{40}Ar^{13+}$ has been performed successfully [23]

10 Summary

Spectroscopy in Penning traps has reached an amazing level of precision even on exotic systems and has opened up many new fields of research. In this chapter we have summarised the results of recent experiments. They provide, to date, the most accurate values of the atomic masses of the electron, proton, and antiproton, the most accurate magnetic moments of the proton and antiproton, and the most stringent test of bound-state quantum electrodynamic calculations of contributions to the magnetic moment of the electron in strong electric fields through g-factor measurements on various hydrogen- and lithium-like highly charged ions.

Acknowledgements We gratefully acknowledge excellent collaboration with Z. Harman, Ch. Keitel, F. Köhler, W. Quint, V. Shabaev, D. Glazov, and S. Sturm. Financial support was provided by the Max-Planck-Gesellschaft, by IMPRS-PTFS and IMPRS-QD, by the European Union (ERC grant AdG 832848—FunI), and by the German Research Foundation (DFG) Collaborative Research Centre "SFB 1225 (ISOQUANT)."

References

1. L.S. Brown, G. Gabrielse, Rev. Mod. Phys. **58**, 233–311 (1986)
2. F.G. Major, V. Gheorghe, G. Werth, *Charged Particle Traps* (Springer, Heidelberg, 2002)
3. K. Blaum, Y.N. Novikov, G. Werth, Contemp. Phys. **51**, 149 (2010)
4. M. Diederich et al., Hyperfine Interact. **115**, 185 (1998)
5. D. Wineland, H. Dehmelt, J. Appl. Phys. **46**, 919 (1975)
6. F. Heiße et al., Phys. Rev. Lett. **119**, 033001 (2017)
7. CODATA Recommended Values of the Fundam. Phys. Const.: 2018NIST SP 961 (2019)
8. S. Ulm et al., Nature **524**, 196 (2015)
9. T. Aoyama et al., Phys. Rev. Lett. **109**, 111807 (2012)
10. D. Hanneke et al., Phys. Rev. A **83**, 052122 (2011)
11. G. Breit, Nature **122**, 649 (1928)
12. V.A. Yerokhin, Z. Harman, Phys. Rev. A **88**, 042502 (2013). and references therein
13. V.A. Yerokhin, C.H. Keitel, Z. Harman, J. Phys. B **46**, 245002 (2013)
14. H. Dehmelt, Proc. Natl. Acad. Sci. U.S.A. **83**, 2291 (1986)
15. N. Hermanspahn et al., Phys. Rev. Lett. **84**, 427 (2000)
16. S. Sturm et al., Phys. Rev. A **87**, 030501(R) (2013)
17. P.J. Mohr, B.N. Taylor, D.B. Newell, CODATA recommended values of the fundamental physical constants. Rev. Mod. Phys. **84**, 1527 (2012)
18. D.A. Glazov et al., Phys. Rev. Lett. **123**, 173001 (2019)
19. H. Häffner et al., Phys. Rev. Lett. **85**, 5308 (2000)
20. J. Verdu et al., Phys. Rev. Lett. **92**, 093002 (2004)
21. F. Köhler et al., Nature Comm. **7**, 10264 (2016)
22. A. Wagner et al., Phys. Rev. Lett. **110**, 033003 (2013)
23. I. Arapoglou et al., Phys. Rev. Lett. **122**, 253001 (2019)
24. G. Schneider et al., Science **358**, 1081 (2017)
25. Ch. Smorra et al., Nature **550**, 371 (2017)
26. S. Sturm et al., Nature **506**, 467 (2014)
27. J.C. López-Urrutia et al., J. Phys. Conf. Ser. **2**, 42 (2004)
28. S. Sturm et al., Eur. Phys. J. **227**, 1425 (2019)

10

Grating Diffraction of Molecular Beams: Present Day Implementations of Otto Stern's Concept

Wieland Schöllkopf

Abstract When Otto Stern embarked on molecular-beam experiments in his new lab at Hamburg University a century ago, one of his interests was to demonstrate the wave-nature of atoms and molecules that had been predicted shortly before by Louis de Broglie. As the effects of diffraction and interference provide conclusive evidence for wave-type behavior, Otto Stern and his coworkers conceived two *matter-wave* diffraction experiments employing their innovative molecular-beam method. The first concept assumed the molecular ray to coherently scatter off a plane ruled grating at grazing incidence conditions, while the second one was based on the coherent scattering from a cleaved crystal surface. The latter concept allowed Stern and his associates to demonstrate the wave behavior of atoms and molecules and to validate de Broglie's formula. The former experiment, however, fell short of providing evidence for diffraction of matter waves. It was not until 2007 that the grating diffraction experiment was retried with a modern molecular-beam apparatus. Fully resolved matter-wave diffraction patterns were observed, confirming the viability of Otto Stern's experimental concept. The correct explanation of the experiment accounts for *quantum reflection*, another wave effect incompatible with the particle picture, which was not foreseen by Stern and his contemporaries.

1 Introduction

The time when Otto Stern and his coworkers at the University of Hamburg were running their pioneering molecular-beam experiments almost a century ago, saw disruptive breakthroughs in quantum physics, experimental and theoretical alike. Among the latter was arguably the work of the french physicist Louis de Broglie on the wave nature of massive particles [1]. He came forward with a rather simple formula for the wavelength λ_{dB} of a *matter wave*, predicting that it equals the product

---

W. Schöllkopf (✉)
Fritz-Haber-Institut der Max-Planck-Gesellschaft, Faradayweg 4-6, 14195 Berlin, Germany
e-mail: wschoell@fhi-berlin.mpg.de

of Heisenberg's constant h and the inverse of the classical momentum $p = mv$ of a particle of mass m and velocity v.

$$\lambda_{dB} = \frac{h}{mv} \tag{1}$$

Given the boldness of the concept of matter waves combined with the simplicity of de Broglie's formula it is not surprising that experimentalists—and a theorist-turned experimentalist such as Otto Stern in particular—must have felt challenged to seek experimental evidence for the existence of de Broglie's matter waves.

Diffraction and interference are unambiguous manifestation of wave-type behavior. As such, Otto Stern and his coworkers conceived two matter-wave diffraction experiments employing their molecular beam method. They had to cope with the fact that, according to Eq. (1), a typical de Broglie wavelength, even for lightweight atoms, at room temperature conditions is in the sub-nanometer regime. Observation of diffraction effects for a wavelength that small requires diffractive optical elements, such as gratings or grids, of a similarly small periodicity. In their article UzM[1] no. 11 *Über die Reflexion von Molekularstrahlen* (Fig. 1) that appeared in Zeitschrift für Physik in 1929 [2] Friedrich Knauer and Otto Stern outlined the two methods they considered promising and they pursued for observing diffraction of a molecular beam. The first one assumes the atoms or molecules to scatter off of a plane ruled (machined) grating at grazing incidence conditions, while the second one was based on the scattering from a cleaved crystal surface.

The latter method was essentially analogous to X-ray diffraction from a crystal lattice, a phenomenon already well known by the mid 1920s. Its first observation by Max von Laue, Walter Friedrich, and Paul Knipping in 1912 provided conclusive evidence for both the wave nature of X-rays and the periodic structure of a crystal lattice (for an historical account on von Laue's experiment see Ref. [3]). X-ray wavelengths are of the same order of magnitude as typical de Broglie wavelengths of light atoms at thermal energies. Crystal lattices, with sub-nanometer periodicity, present a natural match to these wavelengths resulting in comparatively large diffraction angles. Obviously, the main difference is that molecular beams, unlike X-rays, cannot penetrate a crystal. Thus, Otto Stern's approach relies on reflection of a molecular beam from a cleaved crystal surface, which needs to be clean and well-ordered on the atomic scale to allow for coherent scattering of atoms or molecules. While Knauer and Stern where able to observe specular reflection from a crystal surface, they could not present convincing evidence for diffraction of the molecular beam by the periodic crystal lattice. It was several months later, after some improvement of the experimental setup, that Otto Stern together with Immanuel Estermann was able to present unambiguous evidence for diffraction [4]. This work from Otto Stern's molecular beam lab provided the first definite evidence for matter-wave behavior of

[1]UzM stands for *Untersuchungen zur Molekularstrahlmethode*, the series of publications from Stern's molecular-beam lab in Hamburg termed *Investigations by the Molecular Ray Method*, c.f. *Otto Stern's Molecular Beam Method and its Impact on Quantum Physics* by Bretislav Friedrich and Horst Schmidt-Böcking in this volume.

(Untersuchungen zur Molekularstrahlmethode aus dem Institut für physikalische Chemie der Hamburgischen Universität. Nr. 11.)

Über die Reflexion von Molekularstrahlen*.

Von **F. Knauer** und **O. Stern** in Hamburg.

Mit 7 Abbildungen. (Eingegangen am 24. Dezember 1928.)

Molekularstrahlen aus H_2 und He werden an hochpolierten Flächen bei nahezu streifendem Einfall spiegelnd reflektiert. Das Verhalten des Reflexionsvermögens ist in Übereinstimmung mit der de Broglieschen Wellentheorie. Die Versuche, Beugung an Strichgittern nachzuweisen, gaben noch kein Resultat. Auch bei Steinsalzspaltflächen wurde (bei steilerem Einfall) spiegelnde Reflexion gefunden. Die an Kristallspaltflächen beobachteten Erscheinungen sind wahrscheinlich als Beugung aufzufassen, wenngleich ihre vollständige Deutung noch aussteht.

Fig. 1 Front page of the UzM paper no. 11 "*On the reflection of molecular beams*" by Friedrich Knauer and Otto Stern as it appeared in Zeitschrift für Physik in 1929 (received by the journal on December 24, 1928) [2]. In English the abstract states: *Molecular beams of H_2 und He are specularly reflected from highly polished surfaces under near grazing incidence. The reflectivity behavior is in agreement with de Broglie's wave theory. Attempts to find evidence for diffraction from a ruled grating have not yet yielded results. For cleaved surfaces of rock salt (at steeper incidence) reflection was found as well. The phenomena observed with cleaved crystal surfaces are likely due to diffraction, albeit a complete interpretation is still missing*

atoms and molecules. In addition, Estermann and Stern were able to quantitatively check and validate Louis de Broglie's wavelength formula, Eq. (1), for atoms and molecules [4–6].

The success of Knauer's and Stern's second method leaves us with the question: What about the first concept they had conceived to demonstrate matter-wave diffraction with molecular beams; diffraction of a molecular beam reflected from a machined line grating under grazing incidence conditions? The remainder of this contribution will focus on this experiment. In Chap. 2 we will review how Knauer and Stern designed and implemented the grating diffraction experiment and see what results they got with their 1928 molecular beam apparatus. In Chap. 3 we will describe the modern implementation of the experiment, and we will discuss the explanation of the results accounting for quantum reflection. Quantum reflection from the attractive long-range branch of the atom–surface interaction potential is another quantum-wave phenomenon which is incompatible with a particle description. Quantum reflection was not foreseen by 1928, although it is a direct consequence of and evidence for quantum-wave behavior just as diffraction is.

2 The Grating Diffraction Experiment by Knauer and Stern in 1928

In the first paragraph of the UzM no. 11 article Knauer and Stern describe their considerations regarding the two experimental methods they are pursuing with the aim to observe matter-wave diffraction. The original German text reads:

> Der Nachweis der Wellennatur schien uns am bequemsten mit einem Gitter zu führen zu sein. Am nächsten läge es, an die bei den Röntgenstrahlen mit so großem Erfolg benutzten Kristallgitter zu denken. Doch ist von vornherein schwer zu übersehen, ob hier Reflexion und Beugung auftreten werden, weil es im Gegensatz zu den Röntgenstrahlen bei den de Broglie-Wellen auf den Potentialverlauf an der äußeren Grenze der Kristalloberfläche ankommt. Wir beabsichtigen deshalb, optische Strichgitter zu benutzen. Hierfür ist die Voraussetzung, dass es zunächst gelingt, Molekularstrahlen spiegelnd zu reflektieren.

For convenience we are here providing an English translation to the unfortunate reader who is not proficient in German.

> Providing evidence of the wave nature appeared to be most straightforward by using a grating. The most obvious choice would be a crystal lattice which has been used to great success with X-rays. It is, however, difficult to predict if reflection and diffraction will occur, because for de Broglie waves, in contrast to X-rays, the shape of the interaction potential at the outer limit of the crystal surface matters. That is why we intend to use optical ruled gratings. It is prerequisite to first succeed in observing specular reflection of molecular beams.

Apparently, Knauer and Stern were not sure if and to what extent the interaction potential that an atom is exposed to at a crystal surface would possibly impede the observation of diffraction. Therefore, they intended to (also) employ optical ruled gratings in their quest for matter-wave diffraction. For this approach to work it is prerequisite to achieve mirror-like (specular) reflection.

In the following lines Knauer and Stern describe why grazing incidence represents a pivotal aspect of the experimental design. Firstly, the grazing incidence geometry allows for the use of a large (macroscopic) grating period length in the range of 0.01–0.1 mm to diffract wavelengths as small as 0.1 nm. At grazing incidence the effective period that determines the diffraction angles is given by the projection of the grating period along the direction of motion of the incoming molecular beam. Thus, for a grazing angle of 1 mrad the effective period of a 10-micron-period grating is as small as 10 nm. While this is still roughly a factor of 100 larger than the de Broglie wavelengths we are dealing with, it results in diffraction angles of several milliradian, which are well within the experimental resolution. Secondly, the surface roughness, which Knauer and Stern estimate to be on the order of 10–100 nm for their well polished surfaces, would prevent mirror-like reflection at steep incidence. However, as it is known from optics, even a rough surface becomes highly reflective at grazing incidence conditions. Knauer and Stern assume that the roughness times the sine of the incidence angle needs to be smaller than the de Broglie wavelengths if one wants to get good reflectivity. That is why they expect mirror-like reflection of molecular beams from their surfaces for milliradian incidence angles. Interestingly, in this consideration Knauer and Stern ignore a possible influence of the atom-surface interaction potential on the reflectivity.

2.1 Apparatus

The apparatus used by Knauer and Stern was already described in some detail in the preceding paper UzM no. 10 [7], which appeared back to back with their UzM no. 11 paper in volume 53 of Zeitschrift für Physik in 1929. The apparatus allowed them to generate molecular beams of various gases including He, H_2, Ne, Ar, CO_2. As can be seen in the original schematics replotted in Fig. 2, the main components included (i) the molecular beam source, (ii) the detector, (iii) the 3-slits beam collimation system, and (iv) the encasing vacuum system. It appears that all four components were state-of-the-art at that time representing significant improvements compared to previously used molecular beam setups.

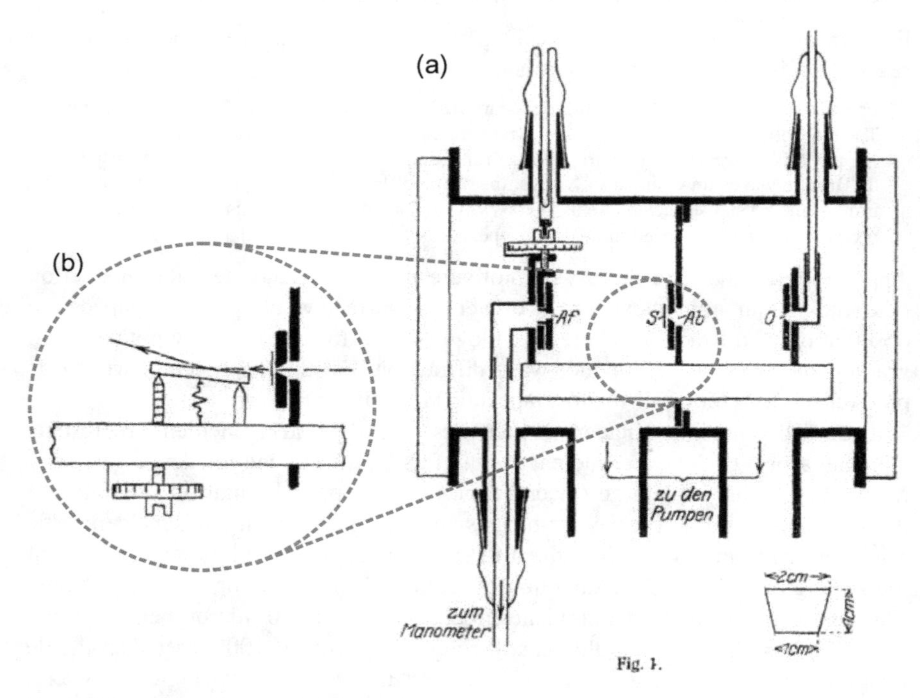

Fig. 2 Schematic of the experimental arrangement used by Knauer and Stern in 1928 copied from the original articles UzM no. 10 and 11. The dashed red circles indicate where the diffraction grating shown in the enlargement (**b**) (from Ref. [2]) was mounted in the molecular-beam apparatus shown in (**a**) (from Ref. [7]). The drawing in the lower right shows the trapezoidal cross section of the horizontal precision nickel-steel rod carrying the slits and the grating mount. The labels *Af*, *Ab*, and *O* denote the three collimating slits; *Auffängerspalt* (collector slit), *Abbildespalt* (imaging slit), and *Ofenspalt* (oven slit i.e. source slit). A shutter (*S*) is located downstream of the imaging slit. As described in the original text, a fourth slit *Vorspalt* (ante slit), not shown in the figure, was used in the grating diffraction experiments. It was located in between *O* and *Ab* and served as a differential pumping stage upstream of *Ab*

Molecular beam source: The gas was fed into a beam source from which it escaped through a narrow slit (10 to 20 μm wide) into the surrounding vacuum chamber. In today's jargon the source would be referred to as an effusive source, where the molecular beam inherits its velocity distribution from the Maxwell-Boltzmann distribution of the gas upstream the slit. Thus, unlike the supersonic free-jet expansion sources that have been available for the last few decades with their inherently narrow velocity distribution, the beams of Knauer and Stern were characterised by relatively large velocity spreads. This, via Eq. (1), corresponds to a wide distribution of de Broglie wavelengths. In other words, Knauer's and Stern's source did not generate monochromatic matter waves. Nonetheless, they were able to adjust the mean velocity and hence the mean de Broglie wavelength by as much as 50% by heating the source or by cooling it down to about 130 K.

Beam collimation and vacuum system: As in classical optics, matter-wave diffraction can only be observed, if spatial coherence is achieved in the experimental setup. This is done by three collimating slits: the first one defines the source (source slit described above); the second one limits the divergence of the beam; and the third one defines the angular resolution of the detector (detector slit described below). All three slits were 10–20 μm wide and 1 cm in height. From the schematic shown in Fig. 2b it can be seen that the second slit also served to effectively separate the vacuum of the right source chamber from the left beam chamber. The chambers were evacuated to a base vacuum pressure of 10^{-5} Torr by two mercury diffusion pumps made by Leybold [7].

Detector: In the Stern Gerlach experiment a few years earlier detection of the beam of silver atoms was accomplished by depositing the beam on a glass plate. After running the experiment for some time, one would check the thickness and location of the deposits. While this method allowed for the detection of the famous two-spot pattern in the Stern Gerlach experiment, it did not enable reliable quantitative measurements of beam intensities, not to mention the fact that the technique did not work with molecular beams of gases. Knauer's and Stern's detector, which is described in the UzM paper no. 10 [7], represents an enormous improvement. Their detector is essentially a Pitot tube with a narrow entrance slit (10–20 μm wide, just as the source slit). The stagnation pressure building up in the detector is measured by a modified Pirani-type vacuum gauge that Knauer and Stern were running at liquid nitrogen temperature. They were able to achieve an impressive absolute sensitivity on the order of 10^{-8} Torr [7]. With the base vacuum pressure in the 10^{-5} Torr regime, this translates into a relative sensitivity of 10^{-3} providing the required sensitivity for the diffraction experiment.

The optical elements—mirrors and gratings: Knauer and Stern were employing different materials for the mirrors they used in the reflectivity measurements: glass, steel, and speculum metal. The text states that each mirror was most thoroughly polished. The ruled gratings they employed in the diffraction experiments were all made from speculum metal with different periods of 10, 20 and 40 μm. No information is provided on how the ruling of the grating was done or what the groove shape might have been.

2.2 Results

Knauer and Stern observed specular reflection of He and H_2 beams from flat solid surfaces made out of any of the three materials. They found the reflectivity to slightly decrease from speculum metal to glass to steel, an observation they explained as a result of the somewhat different surface qualities achievable by polishing these materials. In addition, for each mirror they observed a strong decrease of reflectivity with increasing incidence angle. As an example they provide a reflectivity table for a beam of H_2 reflected from a speculum metal mirror including four data points decreasing from 5% at an incidence angle of 1 mrad to 0.75% at 2.25 mrad. Furthermore, they were able to increase the mean de Broglie wavelength of the molecular beam by cooling the beam source to $-150\,°C$. They state that the observed reflectivity increased by a factor of 1.5 when the mean de Broglie wavelength of the molecular beam was increased by 1.5.

Despite the promising observation of mirror-like specular reflection of the molecular beam it was not possible for Knauer and Stern to observe diffraction when they were employing ruled gratings made out of speculum metal. They searched for diffraction signal in the incidence angle range from 0.5 to 3 mrad, but did not find reliable evidence:

> Wiewohl wir mehrmals Andeutungen eines Maximums gefunden zu haben glauben, gelang es uns nicht, sein Vorhandensein sicherzustellen.

> Although we believe to have seen hints of a [diffraction] peak several times, we were not able to verify its occurrence.

They make this clear in the article's abstract where they summarise *Die Versuche, Beugung an Strichgittern nachzuweisen, gaben noch kein Resultat*, which translates to English as *Attempts to find evidence for diffraction from a ruled grating have not yet yielded results*.

Apparently, Otto Stern planned to try matter-wave diffraction from ruled gratings again with an improved apparatus. As it is described on page 782 of UzM no. 11, a pump of "extremely high pumping speed" was already under construction for Stern's lab at the Leybold company. Stern hoped that the more powerful pump could further boost the molecular beam intensity. We do not know if this apparatus upgrade was implemented. By the time Stern had to leave Hamburg, diffraction from a ruled grating had not been observed in his lab. It would have been described, for instance, in the chapter that Robert Frisch and Otto Stern contributed to the Handbuch der Physik in 1933 [6]. In Peter Toennies's contribution *Otto Stern and Wave-Particle Duality* in this volume the reader will find an account of an interview Otto Stern gave in 1961, where he emphasised his interest in the grating diffraction experiment. As we will see in Sect. 3, Otto Stern's experimental approach was conceptually perfectly viable.

2.3 Historical Note on Friedrich Knauer

From a historical point of view it is intriguing to have a look at Friedrich Knauer, Otto Stern's assistant and only coauthor of the article UzM no. 11 on reflection from ruled gratings. The following basic biographical information is available at Knauer's *Wikipedia* entry: Born in Göttingen in 1897, Knauer studied in Göttingen and Hannover after the First Wold War. He was assistant to Robert Wichard Pohl in Göttingen in 1924, when he moved to Hamburg, where he worked as an associate of Otto Stern's until Stern left in 1933. In that year Knauer completed his Habilitation. He stayed at Hamburg University where he was appointed to the rank of associate professor in 1939. During the Second World War he contributed to the German nuclear project, the so-called *Uranverein*. After the war he continued working at the institute at Hamburg University till 1963. He died in 1979.

Unlike Otto Stern and Immanuel Estermann, Friedrich Knauer was not of Jewish decent. Thus, he was not forced to leave his position at the University in 1933. However, it is somewhat surprising that Friedrich Knauer signed the *Vow of Allegiance of the Professors of the German Universities and High-Schools to Adolf Hitler and the National Socialistic State*[2] that was presented to the public in Leipzig on November 11, 1933. Also Wilhelm Lenz was among the signatories. Lenz was director of the theoretical physics institute in Hamburg to which Wolfgang Pauli was associated. According to Isidor Rabi's recollections [8], Lenz must have had close ties to Stern's group (see also Ref. [9]). Why did Knauer and Lenz sign the Nazi allegiance? Was it out of conviction, fear or opportunism? And what does it mean with regard to Knauer's relation to Otto Stern and other group members like Estermann, who were forced to leave their positions just a few months earlier? After about nine years of fruitful collaboration on pioneering molecular beam experiments, wouldn't one expect some solidarity, or empathy at least, towards the former colleagues? While seeking answers to these questions is beyond the scope of this contribution, it appears doubtful that the few publicly available documents could provide hints to what the answers might be.

3 The Modern Implementation of the Knauer-Stern Experiment

From the two novel experimental methods introduced in UzM no. 11 only scattering from a crystal surface lead to the observation of matter-wave diffraction, while the scattering from a ruled grating did not. The former methods was used by Otto Stern an his associates in follow-up experiments and allowed them to observe, for instance, the appearance of anomalous dips in diffraction patterns [10]. This phenomenon was

[2] *Bekenntnis der Professoren an den Universitäten und Hochschulen zu Adolf Hitler und dem nationalsozialistischen Staat* überreicht vom Nationalsozialistischen Lehrerbund Deutschland, Gau Sachsen, 1933, Dresden-A. 1, Zinzendorfstr. 2.

later explained in terms of selective adsorption by Lennard-Jones and Devonshire [11] (see Peter Toennies's account on *Otto Stern and Wave-Particle Duality* in this volume), and it was developed into an important method of measuring atom-surface interaction potentials. Following the pioneering experiments in Stern's lab, helium atom scattering from crystal surfaces became a tool in surface science that has found widespread application in many labs around the word [12].

In contrast to the success story of scattering molecular beams off of crystal surfaces, the grating diffraction method was not further pursuit. Otto Stern might have tried the experiment again, had he been able to continue his work in Hamburg. But other groups working with molecular beams might not have considered the grating diffraction experiment worthwhile a try, because the de Broglie wave of atoms and molecules was now well established. As a result, Knauer's and Stern's grating diffraction experiment was pretty much forgotten with time.

The situation started to change in the 1990s when the new field of *atom optics and atom interferometry* [13, 14] emerged. The group of David Pritchard at MIT (Cambridge, MA, USA) first demonstrated diffraction of a beam of sodium atoms from a transmission grating [15]. The grating was a free-standing nanoscale structure with a period of only 200 nm. Fabrication of structures that small had become possible by enormous advances in micro-fanbrication and lithography techniques. Unlike in Knauer's and Stern's experiment, with a free-standing grating the molecular beam could pass through the sub-micron slits with no scattering and (almost) no interaction with the grating material. Thus, reflection from a solid surface was no longer prerequisite to observe the grating diffraction pattern. Subsequently, nanoscale transmission gratings were used in a variety of diffraction experiments with molecular beams of atoms, molecules and clusters including; Na_2 [16], metastable He* [17], ground-state He, He_2, and He_3 [18, 19], rare-gas atoms [20], CH_3F and CHF_3 molecules [21]. In addition, Markus Arndt and his coworkers at Vienna University were able to observe diffraction patterns of massive and complex particles starting with C_{60} fullerenes [22] (see Markus Arndt's contribution in this volume). A comprehensive review of transmission-grating diffraction experiments as well as the use of transmission gratings in atom and molecular interferometers can be found in the literature [14].

While diffraction of molecular beams from nanoscale transmission gratings was done by several groups, the original Knauer and Stern experiment was not tried until 2007. In the following we describe the setup of the modern-day implementation of the experiment, the results observed and their interpretation.

3.1 Experimental Setup

The molecular-beam apparatus we used in 2007 is shown schematically in Fig. 3 [23]. The legacy of Otto Stern's molecular-beam experiments is still apparent although, of course, various technical implementations of the beam source, detector, vacuum

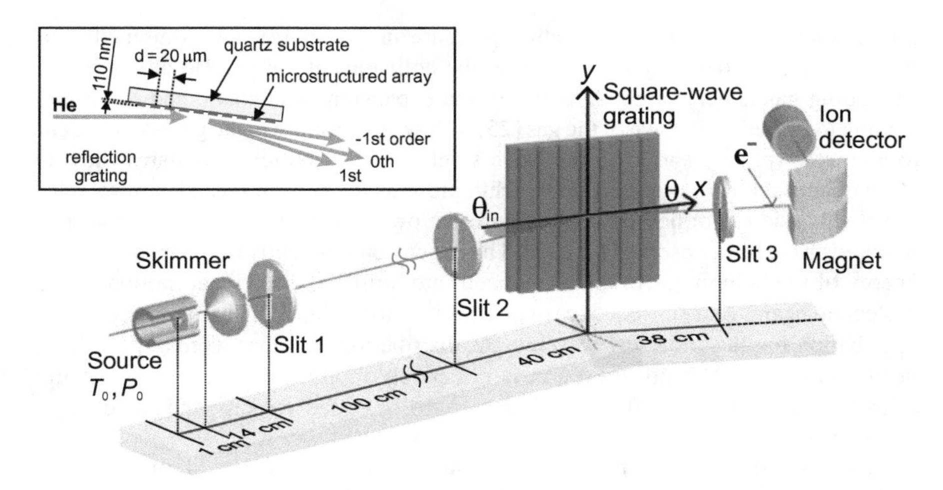

Fig. 3 Schematic of the experimental arrangement used by Zhao et al. in 2007 to observe diffraction of a molecular beam scattering coherently from a reflection grating under grazing incidence conditions [24]. Conceptually, the modern apparatus is essentially analogous to the setup used by Knauer and Stern in 1928, while the technical implementations are different due to, mainly, advances in vacuum and micro-fabrication technologies and in electronics. The inset in the upper left of the figure shows a sketch of the diffraction beams generated at the grating. We use the convention of negative diffraction orders being closer to the grating surface than the specular beam. The incidence angle θ_{in} and detection angle θ are defined with respect to the grating surface plane

system, and the grating are different, not to mention the computerised data acquisition method that was not available to Knauer and Stern.

The beam is formed by free-jet expansion of pure ^{4}He gas from a source cell (stagnation temperature T_0 and pressure P_0) through a 5-μm-diameter orifice into high vacuum. As indicated in Fig. 3, the beam is collimated by two narrow slits, each 20 μm wide, located 15 cm and 115 cm downstream from the source. A third 25-μm-wide detector-entrance slit, located 38 cm downstream from the grating, limits the angular width of the atomic beam to a full width at half maximum of $\approx$ 120 μrad. The detector, which is an electron-impact ionization mass spectrometer, can be rotated precisely around the angle θ indicated in Fig. 3. The reflection grating is positioned at the intersection of the horizontal atom beam axis and the vertical detector pivot axis such that the incident beam approaches under grazing incidence (incident grazing angle $\theta_{in} \leq 20$ mrad), with the grating lines oriented parallel to the pivot axis. Diffraction patterns are measured for fixed incidence angle by rotating the detector around θ and measuring the signal at each angular position. In addition, the grating can be removed from the beam path all together making it possible to measure the direct beam profile, i.e. the undisturbed incident beam, as a function of θ.

Comparison with Stern's apparatus reveals two essential differences beyond mere technological advancements. The first one is the modern molecular beam source. As a consequence of the high stagnation pressure in the source cell combined with the very small 5-μm-diameter aperture, the collision rate of the He atoms in the expanding gas is very large. As a result, the expansion is adiabatic and isentropic leading to a rapid cool down of the gas [25]. A low temperature in the gas is equivalent to a narrow velocity spread in the beam's velocity distribution. The narrowness is often quantified by the *speed ratio* which represents the ratio of mean beam velocity to velocity width. For helium, temperatures below 1 mK and speed ratios of several hundreds have been observed [26, 27]. The expansion efficiently transfers the kinetic energy of the helium gas in the source cell into uniform, directional motion of the molecular beam. As a consequence of de Broglie's formula, Eq. (1), a narrow velocity distribution implies a narrow wavelengths distribution. In other words, the helium beam generated by the modern source is, effectively, monochromatic. In contrast, the effusive beams in the 1920s were characterised by broad velocity and wavelength distributions.

A second qualitative improvement common in most if not all modern-day molecular-beam apparatus is the ionization detector. The neutral He atoms entering the detector are first ionized in collision with electrons of $\simeq 100$ eV kinetic energy. The ions are accelerated by high voltage, mass selected in a magnetic field and finally efficiently detected using a multiplier. The bottle neck of this detection scheme is the inefficient ionization step. It is assumed that only 10^{-6} to 10^{-5} of the neutral He atoms are ionized. However, this poor detection efficiency allows for a far better sensitivity than the Pitot-tube detection scheme that was available to Knauer and Stern in 1928. Interestingly, in an interview given to John L. Heilbron in December 1962 Immanuel Estermann describes that he was trying to build an electron-impact ionization detector in Stern's lab in the early 1920s [28]:

> One of the big problems in molecular beam technology was the (detector) end; we couldn't detect very well. (...) Then, I think, the great step forward was the Langmuir-Taylor detector which was worked out when Taylor was a visiting scientist in Hamburg. But that works only for a limited number of elements or substances. Then we tried all kinds of things, and one of the things that I tried is what is now known as the cross-fire method; it means to bombard the neutral atoms with electrons, thus ionizing them, and then collect the ions. Ions are far easier to detect than the neutral particles. But I did not succeed; I did not get it to work. It's a method which is now used in a number of places quite successfully, but it required much better vacuum technology and much better electronic technology than was available in those days. This must have been about 1923 or '24 when I tried this.

The reflection grating used in 2007 consists of a 56-mm-long micro-structured array of 110-nm-thick, 10-μm-wide, and 5-mm-long parallel chromium strips on a flat quartz substrate. It was made from a commercial chromium mask blank by e-beam lithography. As shown in the inset of Fig. 3 the center-to-center distance of the strips, and thereby the period d, is 20 μm. Given this geometry the quartz surface between the strips is completely shadowed by the strips for all the incidence angles used. We expect a chromium oxide surface to have formed while the grating was exposed to air before mounting it in the apparatus where the ambient vacuum is about 8×10^{-7} mbar. No in-situ surface preparation was done.

3.2 Results

Figure 4 shows a series of diffraction patterns measured with the source kept at $T_0 = 20$ K corresponding to a de Broglie wavelength of $\lambda_{dB} = 2.2$ Å. The incident grazing angle θ_{in} was varied from 3 up to 15 mrad. In each diffraction pattern the specular reflection (0th diffraction-order peak) appears as the strongest peak, for which the detection angle is equal to the incident grazing angle. The intensity of the specular peak decreases continuously from about 600 counts/s at $\theta_{in} = 3.1$ mrad to only 13 counts/s at $\theta_{in} = 15.2$ mrad. At $\theta_{in} = 3.1$ mrad at least seven positive-order diffraction peaks can be seen at angles larger than the specular angle (diffraction 'away from' the surface).

It is straightforward to calculate the nth-order diffraction angle θ_n for given incidence angle θ_{in}, grating period d, and de Broglie wavelength λ_{dB} from the grating equation $\cos(\theta_{in}) - \cos(\theta_n) = n\frac{\lambda_{dB}}{d}$ well known from classical optics [29]. The calculated diffraction angles agree with the observed ones within the experimental error confirming the interpretation of the peaks as grating-diffraction peaks [24]. Note that with increasing incidence angle the negative-order diffraction peaks appear succes-

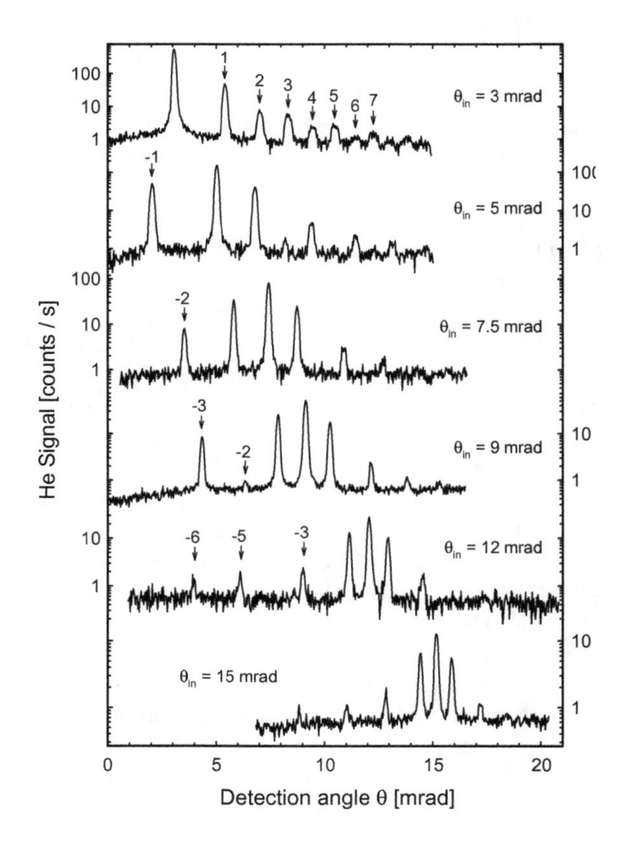

Fig. 4 Diffraction patterns observed for He atom beams of $\lambda_{dB} = 2.2$ Å de Broglie wavelength scattered from a plane grating with 20 μm period at various incidence angles from 3 to 15 mrad (from Ref. [24]). Numbers indicate the diffraction-order assigned to the peaks. In each spectrum the specular peak is most intense. Its peak height decays rapidly with increasing incidence angle

sively, emerging from the grating surface. Emergence of a new diffraction beam comes along with an abrupt redistribution of the flux among the diffraction peaks. These emerging-beam resonances have been studied for He atom beams diffracted by a blazed ruled grating [30].

The relative diffraction peak intensities change significantly with incident grazing angle. For instance, for $\theta_{in} = 3.1$ mrad even and odd order peaks have similar heights falling off almost monotonously with increasing diffraction order. With increasing incident grazing angle, however, the positive even-order diffraction peaks tend to disappear. Moreover, a distinct peak-height variation can be seen for the -2nd-order peak which decreases sharply when θ_{in} is increased from 7.4 to 9.1 mrad.

A diffraction pattern as the ones in Fig. 4 must have been exactly what Otto Stern was longing to see. The data shown here demonstrates that Stern's concept was perfectly viable. What prevented Knauer and Stern from observing diffraction with a machined grating was the limited experimental technology of their time that did not provide the required high beam flux, efficient detection, and high vacuum conditions.

3.3 Quantum Reflection

The peak heights decaying rapidly with increasing incidence angle confirm Otto Stern's conjecture that the highest reflectivity is to be found for the most grazing incidence conditions. A quantitative analysis of the reflectivity dependence on incidence angle is shown in Fig. 5. The reflectivity was determined by summing up the areas of all the peaks in a diffraction pattern. The sum was divided by the area of the direct beam to give the reflectivity. The direct beam (not shown in the plots) is measured when the grating is completely removed from the beam path. The reflectivity at various incidence angles and source temperatures is plotted in Fig. 5 in a semi-logarithmic plot as a function of the incident atoms' wave-vector component perpendicular to the surface, k_{perp}. It is apparent that all the data points fall on a single curve. At small perpendicular wave-vector up to about $0.11\,\mathrm{nm}^{-1}$ the curve decays rapidly, while at values larger than about $0.13\,\mathrm{nm}^{-1}$ it decays at reduced rate.

The same characteristic reflectivity behavior is also found when the diffraction grating is replaced by a plane surface [31]. This behavior cannot be understood by considering a wave scattering off of a rough surface, if one does not include the subtle effects of the atom–surface interaction. The expected dependence in the absence of a surface potential is also plotted in Fig. 5. While the reflectivity in the non-interaction model also tends to unity in the limit of vanishing perpendicular wave vector, this model obviously fails to describe the reflectivity dependence on k_{perp}.

A decent agreement with the observed data is found when quantum reflection from the atom–surface interaction potential is considered. Quantum reflection, just as diffraction, is a wave effect not compatible with the particle picture. The basic idea behind quantum reflection is illustrated in Fig. 6a, b, where the atom–surface interaction potential $V(z)$ is approximated by a square well potential, i.e., $V(z) = \infty$

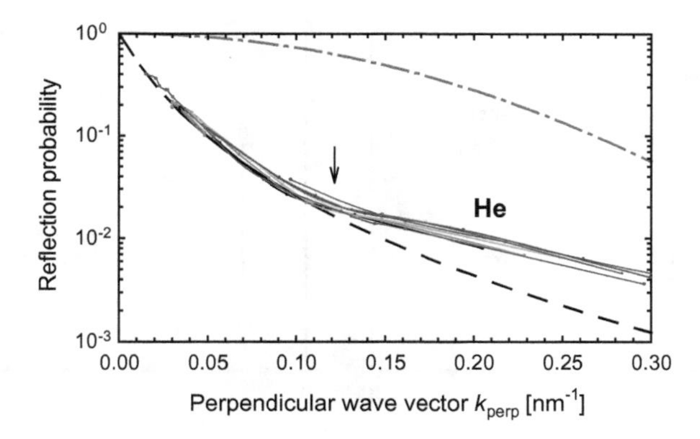

Fig. 5 Reflection probability of He atoms of various de Broglie wavelengths scattering from a plane grating with 20 μm period at a variety of grazing incidence angles plotted as a function of the normal component of the incident atom's wave vector [24]. Each color corresponds to a different wavelength. The dashed line presents a 1-dimensional quantum reflection calculation fitted to the data points at small perpendicular wave vector (left of the arrow). The dash-dotted line represents the reflection probability calculated for a wave scattering from a rough surface (4 nm root-mean-square roughness) in the absence of an atom–surface interaction potential

for $z < a$, $V(z) = 0$ for $z > b$, and $V(z) = V_0$, where V_0 is negative, for $a \leq z \leq b$. Here, the variable z denotes the atom to surface distance, and a and b are positive constants.

Within the classical particle description, Fig. 6a, an incident He atom approaching the well region from positive z with initial kinetic energy E_{kin} will gain energy upon entering the well at $z = b$; its kinetic energy is increased by exactly the well depth V_0. With correspondingly larger velocity the particle slams onto the steep repulsive wall where it is scattered back at the classical turning point. Upon leaving the well its kinetic energy is reduced to its initial value.

The description looks different in the quantum-wave picture, Fig. 6b, if the initial kinetic energy E_{kin} is sufficiently small such that quantum effects become observable. We then have to deal with a wave approaching a step in the potential at $z = b$. As quantum mechanics teaches us, in this situation there is a non-vanishing probability for reflection at the step (even for a "step down" as in our system). This quantum-wave reflection probability increases with increasing de Broglie wavelength of the incident atom. For a discontinuous step, as the one shown in Fig. 6b, it even approaches unity in the limit of vanishing kinetic energy.

The square-well model is a simplistic approximation to an atom–surface potential. In a more realistic model the steps will necessarily be smoothed out, as indicated in the depiction shown in Fig. 6c. Even then, there will be an appreciable reflection probability at the attractive branch of the potential as long as the incident energy of the atom is sufficiently small. This reflection mechanism of matter waves, referred to as quantum reflection, occurs in absence of a classical turning point and, paradoxically,

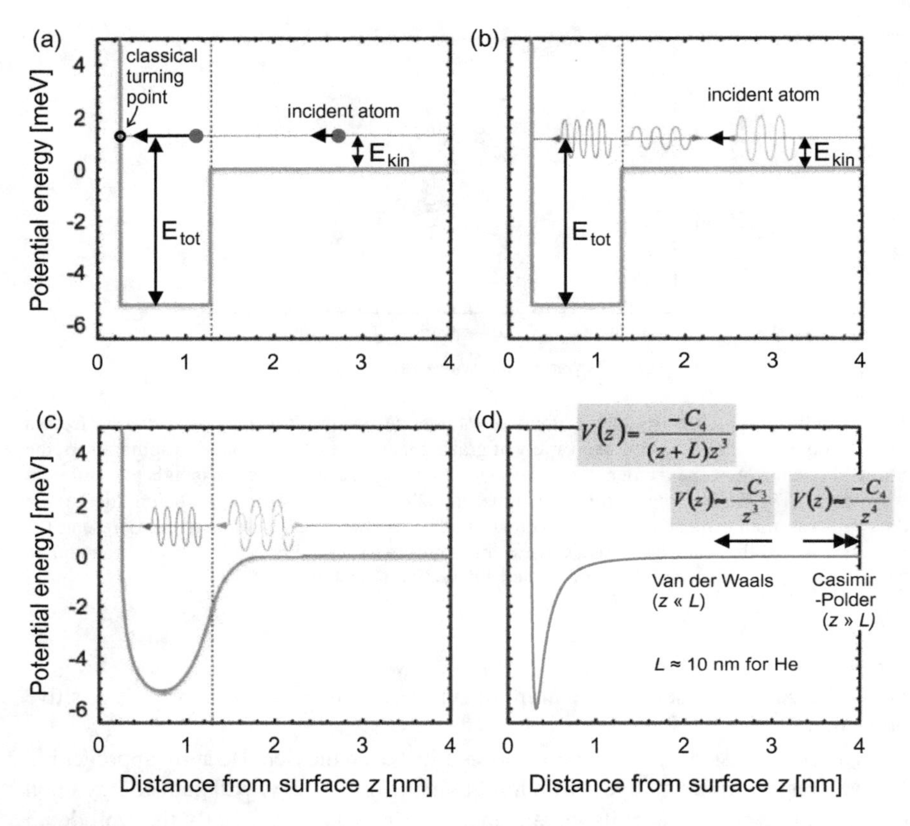

Fig. 6 Illustration of quantum reflection of an atom from a solid surface. In **a** and **b** the atom–surface interaction potential is approximated by a square-well model combined with the classical particle picture (**a**) and the quantum wave picture (**b**). In **c** the square well is smoothened, while **d** shows the potential between a He atom and a silver surface as an actual example. The long-range attractive branch of the potential exhibits a transition from the van der Waals regime at intermedium *z* to the Casimir–Polder regime at very large *z*

in the absence of a repulsive force acting on the incoming atom. The latter aspect in particular is counter-intuitive and incompatible with the classical description.

A realistic atom–surface potential model is displayed in Fig. 6d. The example shown describes the interaction between a He atom and a silver surface with a well depth of about 6 meV. The attractive potential branch is modelled by the function $V(z) = \frac{-C_4}{(z+L)z^3}$ describing a Van der Waals–Casimir interaction. The atom specific length L ($L \approx 10$ nm for He) marks the transition from the Van der Waals regime ($V(z) \propto z^{-3}$) at $z \ll L$ to the Casimir–Polder regime ($V(z) \propto z^{-4}$) at $z \gg L$. Although this potential looks smooth, it can be shown that its quantum reflection probability will approach unity in the zero-energy limit [32].

Quantum reflection of atoms from a solid surface was described by theory [32–38]. It was first observed in experiments with ultracold metastable atoms [39] and, later, also with a Bose-Einstein condensate [40, 41]. In these experiments extremely small atomic velocities needed to observe quantum reflection are achieved by cooling a dilute atomic gas in a trap to ultracold temperatures. An alternative approach to achieve those velocities is to scatter an atomic or molecular beam from a solid at grazing incidence [24, 31, 42], Due to the grazing incidence geometry the relevant velocity z-component, perpendicular to the surface, can approach extremely small values allowing observation of quantum reflection. The comparatively large parallel velocity component does not affect the quantum reflection process as long as the surface is (at least locally) homogenous. Quantum reflection of helium beams from plane surfaces [31, 42] as well as laminar [24] and blazed ruled [30] gratings has been reported.

In addition, reflection and diffraction from a grating was also observed for weakly bound ground-state helium dimers (He_2 binding energy ≈ 0.1 μeV) and trimers (He_3 binding energy ≈ 10 μeV) [43, 44]. Following the above description of classical scattering, the forces in the molecule–surface potential well region will inevitably lead to bond breakup, because the well depth (order of magnitude 10 meV) is $\approx 10^5$ times and $\approx 10^3$ times larger than the binding energy of helium dimers and trimers, respectively. Therefore, the observation of reflection of dimers and trimers provides direct evidence for quantum reflection. Furthermore, the fact that diffraction patterns are found indicates that quantum reflection leads to coherent reflection of matter waves.

4 Conclusion

The experiment described by Friedrich Knauer and Otto Stern in the UzM no. 11 article was designed to observe matter-wave diffraction of He and H_2 beams scattering off of a ruled diffraction grating at grazing incidence conditions. They were able to observe mirror-like, specular reflection with the reflectivity increasing with decreasing incidence angle and with increasing de Broglie wavelength. But, they fell short of detecting diffraction peaks. The modern reincarnation of the experiment was performed in 2007 in the Fritz-Haber-Institut der Max-Planck-Gesellschaft in Berlin, Germany. Conceptually, the modern apparatus is in line with the setup used in Otto Stern's lab. Yet, its intense molecular-beam source and sensitive electron-impact ionization detector out-perform their 1920s counterparts. In combination with better vacuum systems and modern electronics this made it possible to observe fully resolved diffraction patterns of He atom beams including peaks up to the seventh diffraction order. It can be assumed that the diffraction pattern observed in 2007 was exactly what Knauer and Stern were hoping to observe.

The results gained with the modern equipment make it clear that Knauer's and Stern's experiment was well conceived; it was bound to work, as soon as sufficient signal was available. This holds despite the fact that Otto Stern and his coworkers did not know that the atom–surface interactions can lead to quantum reflection and, thus, play a crucial role in the coherent scattering of atoms and molecules from surfaces.

While they did consider atom–surface interactions in the context of diffraction from a cleaved crystal surface, they ignored it for the gratings. It might well be that Friedrich Knauer and Otto Stern were the first to observe, unknowingly, quantum reflection of atoms from a solid surface more than 80 years before the first conclusive demonstration of this effect by Shimizu [39]. Ironically, quantum reflection itself is a wave phenomenon that cannot be explained with particles. As such, observation of quantum reflection of atoms provides evidence for de Broglie waves of an atom, exactly what Otto Stern was aiming to observe.

Acknowledgements I am indebted to my long-term collaborator, Prof. Bum Suk Zhao (UNIST, Ulsan, South Korea). His hard and skilful work was crucial to succeed with our joint diffraction experiments described in this contribution. I also want to thank Prof. Peter Tonnies (Max-Planck-Institut für Dynamik und Selbstorganisation, Göttingen, Germany). As my thesis adviser he taught me the art of running molecular-beam experiments; the legacy from Otto Stern a century ago. Furthermore, I thank Prof. Gerard Meijer (Fritz-Haber-Institut (FHI) der Max-Planck-Gesellschaft, Berlin, Germany) for making it possible for me to pursue molecular-beam diffraction experiments at the FHI. Last not least, I thank Prof. Bretislav Friedrich (FHI, Berlin, Germany) for numerous revealing discussions on Otto Stern and his legacy, and I want to thank him as well as Prof. Horst Schmidt-Böcking (Goethe Universität, Frankfurt, Germany) for their effort of organising this one-of-a-kind Otto-Stern tribute.

References

1. L. de Broglie, Recherches sur la théorie des quanta. Annales de Physique **10**, 22–128 (1925)
2. F. Knauer, O. Stern, The Reflection of Molecular Beams. Zeitschrift für Physik **53**, 779–791 (1929)
3. M. Eckert, Max von Laue and the discovery of X-ray diffraction in 1912. Ann. Phys. **524**, A83–A85 (2012)
4. I. Estermann, O. Stern, Diffraction of molecular beams. Zeitschrift für Physik **61**, 95–125 (1930)
5. I. Estermann, R. Frisch, O. Stern, Molecular ray problems. Experiments with monochromatic de Broglie waves of molecular beams. Physikalische Zeitschrift **32**, 670–674 (1931)
6. O.R. Frisch, O. Stern, *Handbuch der Physik*, ed. by H. Geiger, K. Scheel, Vol. XXII, II. Teil (Negative und positive Strahlen), 2 edn. (Springer, Berlin, Germany, 1933) , pp. 313–354
7. F. Knauer, O. Stern, Intensity measurements on molecular beams of gases. Zeitschrift für Physik **53**, 766–778 (1929)
8. T.S. Kuhn, Interview of I.I. Rabi (1963). Available online at www.aip.org/history-programs/niels-bohr-library/oral-histories/4836
9. J.P. Toennies, H. Schmidt-Böcking, B. Friedrich, J.C.A. Lower, Otto Stern (1888–1969): the founding father of experimental atomic physics. Ann. Phys. (Berlin) **523**, 1045–1070 (2011)
10. R. Frisch, O. Stern, Abnormality in the specular reflection and diffraction of molecular beams of crystal cleavage planes. I. Zeitschrift für Physik **84**, 430–442 (1933)
11. J.E. Lennard-Jones, A.F. Devonshire, Diffraction and selective adsorption of atoms at crystal surfaces. Nature **137**, 1069–1070 (1936)
12. G. Benedek, J.P. Toennies, *Atomic Scale Dynamics at Surfaces: Theory and Experimental Studies with Helium Atom Scattering* (Springer, Berlin, Germany, 2018)
13. P.R. Berman (ed.), *Atom Interferometry* (Academic Press, New York, 1997)
14. A.D. Cronin, J. Schmiedmayer, D.E. Pritchard, Optics and interferometry with atoms and molecules. Rev. Mod. Phys. **81**, 1051–1129 (2009)

15. D.W. Keith, M.L. Schattenburg, H.I. Smith, D.E. Pritchard, Diffraction of atoms by a transmission grating. Phys. Rev. Lett. **61**, 1580 (1988)
16. M.S. Chapman, et al., Optics and interferometry with Na_2 molecules. Phys. Rev. Lett. **74**, 4783 (1995)
17. O. Carnal, A. Faulstich, J. Mlynek, Diffraction of metastable helium atoms by a transmission grating. Appl. Phys. B **53**, 88 (1991)
18. W. Schöllkopf, J.P. Toennies, Nondestructive mass selection of small van der Waals clusters. Science **266**, 1345 (1994)
19. W. Schöllkopf, J.P. Toennies, The nondestructive detection of the helium dimer and trimer. J. Chem. Phys. **104**, 1155 (1996)
20. R.E. Grisenti, et al., Determination of atom-surface van der Waals potentials from transmission-grating diffraction intensities. Phys. Rev. Lett. **83**, 1755–1758 (1999)
21. W. Schöllkopf, R.E. Grisenti, J.P. Toennies, Time-of-flight resolved transmission-grating diffraction of molecular beams. Eur. Phys. J. D **28**, 125 (2004)
22. M. Arndt, O. Nairz, J. Vos-Andreae, C. Keller, G. van der Zouw, A. Zeilinger, Wave-particle duality of C_{60} molecules. Nature **401**, 680 (1999)
23. This apparatus was built in the 1990s in Prof. J.P. Toennies's Dept. of Molecular Interactions in the Max-Planck-Institut für Strömungsforschung in Göttingen, Germany, and was relocated to the Fritz-Haber-Institut der Max-Planck-Gesellschaft in Berlin, Germany in 2005 where it has been in use since then
24. B.S. Zhao, S.A. Schulz, S.A. Meek, G. Meijer, W. Schöllkopf, Quantum reflection of helium atom beams from a microstructured grating. Phys. Rev. A **78**, 010902(R) (2008)
25. H. Buchenau, E.L. Knuth, J. Northby, J.P. Toennies, C. Winkler, Mass spectra and time-of-flight distributions of helium cluster beams. J. Chem. Phys. **92**, 6875 (1990)
26. J. Wang, V.A. Shamamian, B.R. Thomas, J.M. Wilkinson, J. Riley, C.F. Giese, W.R. Gentry, Speed ratios greater than 1000 and temperatures less than 1 mk in a pulsed He beam. Phys. Rev. Lett. **60**, 696–699 (1988)
27. L.W. Bruch, W. Schöllkopf, J.P. Toennies, The formation of dimers and trimers in free jet ^{4}He cryogenic expansions. J. Chem. Phys. **117**, 1544–1566 (2002)
28. J.L. Heilbron, Interview of Immanuel Estermann (1962). Available online at www.aip.org/history-programs/niels-bohr-library/oral-histories/4593
29. M. Born, E. Wolf, *Principles of Optics*, 6th edn. (Cambridge University Press, Cambridge, 1997)
30. B.S. Zhao, G. Meijer, W. Schöllkopf, Emerging beam resonances in atom diffraction from a reflection grating. Phys. Rev. Lett. **104**, 240404 (2010)
31. B.S. Zhao, H.C. Schewe, G. Meijer, W. Schöllkopf, Coherent reflection of He atom beams from rough surfaces at grazing incidence. Phys. Rev. Lett. **105**, 133203 (2010)
32. H. Friedrich, G. Jacoby, C.G. Meister, Quantum reflection by Casimir-van der Waals potential tails. Phys. Rev. A **65**, 032902 (2002)
33. R.B. Doak, A.V.G. Chizmeshya, Sufficiency conditions for quantum reflection. Europhys. Lett. **51**, 381–387 (2000)
34. A. Mody, M. Haggerty, J.M. Doyle, E.J. Heller, No-sticking effect and quantum reflection in ultracold collisions. Phys. Rev. B **64**, 085418 (2001)
35. S. Miret-Artés, E. Pollak, Scattering of He atoms from a microstructured grating: quantum reflection probabilities and diffraction patterns. J. Phys. Chem. Lett. **8**, 1009–1013 (2017)
36. J. Petersen, E. Pollak, S. Miret-Artés, Quantum threshold reflection is not a consequence of a region of the long-range attractive potential with rapidly varying de Broglie wavelength. Phys. Rev. A **97**, 042102 (2018)
37. G. Rojas-Lorenzo, J. Rubayo-Soneira, S. Miret-Artés, E. Pollak, Quantum reflection of rare-gas atoms and clusters from a grating. Phys. Rev. A **98**, 063604 (2018)
38. G. Rojas-Lorenzo, J. Rubayo-Soneira, S. Miret-Artés, E. Pollak, Quantum threshold reflection of He-atom beams from rough surfaces. Phys. Rev. A **101**, 022506 (2020)
39. F. Shimizu, Specular reflection of very slow metastable neon atoms from a solid surface. Phys. Rev. Lett. **86**, 987–990 (2001)

40. T.A. Pasquini, et al., Quantum reflection from a solid surface at normal incidence. Phys. Rev. Lett. **93**, 223201 (2004)
41. T.A. Pasquini, et al., Low velocity quantum reflection of Bose-Einstein condensates. Phys. Rev. Lett. **97**, 093201 (2006)
42. V. Druzhinina, M. DeKieviet, Experimental observation of quantum reflection far from threshold. Phys. Rev. Lett. **91**, 193202 (2003)
43. B.S. Zhao, G. Meijer, W. Schöllkopf, Quantum reflection of He_2 several nanometers above a grating surface. Science **331**, 892–894 (2011)
44. B.S. Zhao, W. Zhang, W. Schöllkopf, Non-destructive quantum reflection of helium dimers and trimers from a plane ruled grating. Mol. Phys. **111**, 1772–1780 (2013)

Permissions

All chapters in this book were first published by Springer; hereby published with permission under the Creative Commons Attribution License or equivalent. Every chapter published in this book has been scrutinized by our experts. Their significance has been extensively debated. The topics covered herein carry significant information for a comprehensive understanding. They may even be implemented as practical applications or may be referred to as a beginning point for further studies.

The contributors of this book come from diverse backgrounds, making this book a truly international effort. We would like to thank all the contributing authors for lending their expertise to make the book truly unique. They have played a crucial role in the development of this book. Without their invaluable contributions this book wouldn't have been possible. They have made vital efforts to compile up to date information on the varied aspects of this subject to make this book a valuable addition to the collection of many professionals and students.

This book was conceptualized with the vision of imparting up-to-date and integrated information in this field. To ensure the same, a matchless editorial board was set up. Every individual on the board went through rigorous rounds of assessment to prove their worth. After which they invested a large part of their time researching and compiling the most relevant data for our readers.

The editorial board has been involved in producing this book since its inception. They have spent rigorous hours researching and exploring the diverse topics which have resulted in the successful publishing of this book. They have passed on their knowledge of decades through this book. To expedite this challenging task, the publisher supported the team at every step. A small team of assistant editors was also appointed to further simplify the editing procedure and attain best results for the readers.

Apart from the editorial board, the designing team has also invested a significant amount of their time in understanding the subject and creating the most relevant covers. They scrutinized every image to scout for the most suitable representation of the subject and create an appropriate cover for the book.

The publishing team has been an ardent support to the editorial, designing and production team. Their endless efforts to recruit the best for this project, has resulted in the accomplishment of this book. They are a veteran in the field of academics and their pool of knowledge is as vast as their experience in printing. Their expertise and guidance has proved useful at every step. Their uncompromising quality standards have made this book an exceptional effort. Their encouragement from time to time has been an inspiration for everyone.

The publisher and the editorial board hope that this book will prove to be a valuable piece of knowledge for students, practitioners and scholars across the globe.

List of Contributors

Klaas Bergmann
Fachbereich Physik der Technischen Universität Kaiserslautern, Erwin Schrödinger Str. 49, 67663 Kaiserslautern, Germany

J. Peter Toennies
Max-Planck-Institut für Dynamik und Selbstorganisation, Am Fassberg 17, 37077 Göttingen, Germany

Gilbert M. Nathanson
Department of Chemistry, University of Wisconsin-Madison, 1101 University Avenue, Madison, WI 53706, USA

M. S. Schöffler, L. Ph. H. Schmidt, S. Eckart, R. Dörner and T. Jahnke
Institut für Kernphysik, Universität Frankfurt, 60348 Frankfurt, Germany

A. Czasch, O. Jagutzki and H. Schmidt-Böcking
Institut für Kernphysik, Universität Frankfurt, 60348 Frankfurt, Germany
Roentdek GmbH, 65779 Kelkheim, Germany

J. Ullrich
PTB, Brunswick, Germany

R. Moshammer
MPI für Kernphysik, Heidelberg, Germany

R. Schuch
Physics Department, Stockholm University, 107 67 Alba Nova, Stockholm, Sweden

Stefan Gerlich, Yaakov Y. Fein, Armin Shayeghi and Markus Arndt
Faculty of Physics, University of Vienna, Boltzmanngasse 5, 1090 Vienna, Austria

Valentin Köhler and Marcel Mayor
University of Basel, St. Johannsring 19, 4056 Basel, Switzerland

John S. Briggs
Institute of Physics, University of Freiburg, Freiburg, Germany
Department of Physics, Royal University of Phnom Penh, Phnom Penh, Cambodia

Hendrik Ulbricht
School of Physics and Astronomy, University of Southampton, Highfield SO17 1BJ, UK

Christiane P. Koch
Arnimallee 14, 14195 Berlin, Germany

Klaus Blaum
Max-Planck-Institut für Kernphysik, Heidelberg, Germany

Günter Werth
Johannes-Gutenberg-Universität, Mainz, Germany

Wieland Schöllkopf
Fritz-Haber-Institut der Max-Planck-Gesellschaft, Faradayweg 4-6, 14195 Berlin, Germany

Index

Printed in the USA
CPSIA information can be obtained
at www.ICGtesting.com
JSHW011400091023
49903JS00004B/42